THE
CALCULATOR
PUZZLE
BOOK

THE
CALCULATOR
PUZZLE
BOOK

by

CLAUDE BIRTWISTLE

BELL PUBLISHING COMPANY
NEW YORK

Copyright © MCMLXXVIII by Elliot Right Way Books
This edition is published by Bell Publishing Company,
a division of Crown Publishers, Inc.,
by arrangement with Elliot Right Way Books.
a b c d e f g h
BELL 1978 PRINTING
Manufactured in the United States of America

Library of Congress Cataloging in Publication Data

Birtwistle, Claude.
    The calculator puzzle book.

    Includes index.
    1. Calculating-machines—Problems, exercises,
etc.   2. Mathematical recreations.   I. Title.
QA75.B55   793.7′4   78-25783
ISBN 0-517-26682-2

Contents

INTRODUCTION

So you have a pocket calculator, you use it in your work, around the house, when you go shopping and you still find it fascinating to push the buttons and juggle about with numbers? Then here are some puzzles to test your ingenuity and to provide you with an excuse for pushing those buttons many more times!

The puzzles that follow naturally involve a little mathematics, but much of this is fairly simple and the hints explain the mathematics where necessary. There are easy puzzles, some a little more difficult and finally one or two harder ones just to test you! All the puzzles may be solved using a simple four-function calculator.

If you are in difficulties, refer to the *Hints and Answers*, where methods are explained and hints given. Also if you have to look up a hint and feel that you would like to try another puzzle of a similar type, you will find after the answer a reference to similar problems in the book.

So, calculators at the ready? Let us start with some Keyboard Capers.

IMPORTANT NOTE: This book was originally published in Great Britain. The British usage of raised decimal points has been retained throughout. Thus, 2·54 is the same as 2.54. British spellings have also been retained, but pounds and pence have been changed to dollars and cents.

1 KEYBOARD CAPERS

The numbers on the keyboard of an electronic calculator are always arranged like this:

$$7 \quad 8 \quad 9$$
$$4 \quad 5 \quad 6$$
$$1 \quad 2 \quad 3$$

Enter any three figures in a straight line through the centre of this display, e.g. 753. Press the addition key and then enter the same three figures, but this time reading in the opposite direction; in the above example this would be 357. Find the total.

Now repeat the exercise with another line of three numbers; remember you must always go through the centre of the display. Examples of starting numbers may be 159 or 456.

Continue the exercise as long as you can without repeating earlier examples. What do you notice about all the totals? Explain this.

2 MORE KEYBOARD CAPERS

This time the exercise is similar, but instead of adding numbers, we subtract them. Work out these in this order and write down the answers each time:

$$654 - 456 =$$
$$753 - 357 =$$
$$852 - 258 =$$
$$951 - 159 =$$

What do you notice about the answers?
Can you find the reason?

3 GRAND TOTAL

How many different whole numbers can be shown in the display of an electronic calculator with an eight-figure display?

4 IT'S AN UPSIDE DOWN WORLD

Most calculator users know the trick of telling someone with a calculator that last year 28430938 barrels of oil were produced and the profit on each was $0·25. Multiplying these two numbers on the calculator gives the total profit and if you turn the calculator upside down and read the display again, it tells you who made the profit!

The figures are quite fictitious, of course, and before some firm claims damages we could hurriedly change the problem to an output of 2366851 barrels and a profit of 3 dollars a barrel. Turning the display upside down now tells a different story!

Both the examples are well-known, but now try this:

Enter the number 999999 and multiply by 3.

Press the addition key.

Enter the number 6810189, but before you press the equals key to obtain the result of the addition, turn the calculator upside down and look again at the number which you have just entered.

Now press the equals key.

Look at your answer and then turn your calculator upside down and look at the answer again.

9

5 FOOTBALL FANCIES

(*a*) How many different forecasts can be made of the possible results of ten football matches, assuming in each case the equal likelihood of either a home win, an away win or a draw?

(*b*) Repeat the exercise, but this time for the results of twelve football matches.

(*c*) What are the chances of making a correct forecast in each case?

6 TOM'S TEAM

I was at the local soccer ground the other evening and found Tom, the team captain, in the changing room with a blank team sheet and a pencil. 'Puzzling out the team for next Saturday?' I asked.

'There isn't much puzzle,' said Tom. 'We only have eleven players, so there is only one team I can write down!'

'Not so,' said I. 'You have a vast number of possible teams.' Tom looked at me in bewilderment and so I explained.

The players don't have to play always in the same position. So considering the goalkeeper first, there are eleven names you could write on the sheet in that position. For every one of the names there, you could now write any one of the ten remaining names in the next space on the sheet. So for these two places alone you have 11×10, i.e. 110 ways of filling them. Now for every one of these, there are nine ways of filling the next position and so on until the whole team sheet is filled.

Can you say how many different team sheets Tom could write out using only his eleven players?

7 TRAVELLING TRANSACTIONS

A travelling salesman started his journey with a sum of money in his pocket. The first day he doubled the amount of money that he had and then spent $30. The next day he increased his remaining amount of money threefold and then spent $54. On the third day he quadrupled the amount of money he had at the beginning of the day and then spent $72.

At the end of the third day he found that he had just $48 left. How much did he have in his pocket on the first day as he started his journey?

8 RE-ARRANGEMENT

Enter any number into your calculator and subtract from it any smaller number which contains the same digits (arranged in any order). Divide your answer by 9.

For example, 278165 − 187625 = 90540

90540 divided by 9 is 10060.

The final answer does not really matter. The question is whether, when you do this subtraction, the answer is always divisible exactly by 9? If so, why?

9 SAVINGS PLAN

Mary had planned her holiday abroad and she was deciding how to save up the money before she went. It was sixteen weeks before the departure and she reckoned that as the time approached she would become increasingly keen for the holiday and so would be prepared to save more. She decided, therefore, to save $1 the first week, $2 the second week, $3 the third week and so on until in the sixteenth week she would be saving $16. What was the total amount of money that she saved?

It is likely that you added the numbers 1 to 16 in your calculator, but there are easier ways of working out problems like this. Add together the amount saved in the first and last weeks; note the answer. Now add the amount saved in the second and next-to-last weeks. Next add the amount saved in the third and the third-from-end weeks. And so on. What do you notice about all these answers? How many such answers could you have for the whole problem?

Can you now see an easy way to do the calculation? Add the first and last amounts and multiply this sum by half the number of weeks. As an exercise, how much would Mary have saved if she had been saving in this manner for (*a*) 20 weeks, (*b*) 30 weeks?

10 HI-FI

Bill has a Hi-Fi shop and has fitted up a complicated series of wires and switches so that his customers may listen to different combinations of record decks, amplifiers and speakers in order to compare them.

He stocks eight different record decks, twelve different amplifiers and ten different sets of speakers (a set of speakers is a pair and they have to be used simultaneously, of course).

How many different combinations of record deck, amplifier and speakers is a customer able to listen to?

11 GET YOUR TEETH INTO THIS

Three cog-wheels are in mesh. The first one, with seven teeth, drives the second one which has ten teeth. This second wheel in turn drives a third cog-wheel with twelve teeth.

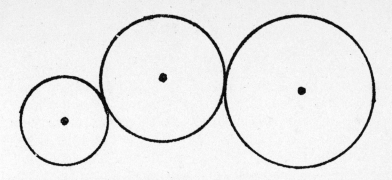

When the wheels are at rest, a mark is made at the point where the first and second wheels are in contact and another mark at the point where the second and third wheels are in contact.

How many times does the first wheel revolve before the marks first return to their original positions together?

12 LOTS OF SQUARES

Use your calculator to find the squares of the numbers from 1 to 9 and write the answers in a column. (The square of a number is obtained by multiplying the number by itself and most calculators will do this by entering the number, then pressing in turn the multiplication and equals keys.)

Next find the squares of numbers from 11 to 19 and write these in a column alongside the first column. Repeat for squares of numbers from 21 to 29.

Look at the results and try to explain:
(i) why the pattern of the last digits is always the same in each column;
(ii) why the first and last numbers in a column have the same last digit, the second and next-to-last numbers have the same last digit, the third and third-from-the-end, and so on.

13 GRAND TOTAL

How many great-great-grandparents did you have, and how many great-great-grandparents did your great-great-grandparents have?

14 FIND AN EASY WAY!

There are four multiples of 3 between 10 and 22. These are 12, 15, 18 and 21 and they are easily found using a calculator with a constant facility. You simply find the multiples by using 3 as the constant and observe the results in the display. The numbers are selected which satisfy the required conditions. In the following examples, however, we do not wish to identify the multiples, but to find how many there are.

How many multiples of 3 are there between 100 and 200?

How many multiples of 7 are there between 50 and 500?

✓15 CROSS NUMBER PUZZLE

Clues across

A $112 \div 256 \times 10000$

E A fifth of the number which is six less than M across

F See L across

H 87^2

J 23×9^2

L E across + F across

M The sum of the first thirteen whole numbers

O $18579624 \times 84 \div 1327116$

Clues down

B The square root of 961

C 6^5

D 16^3

F The sum of the first six odd numbers

G $32^2 + 11^2$

I $599 \times 2484 \div 276$

K 30 per cent of 290

N $3^4 - 4^3$

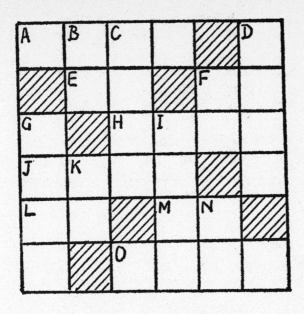

16 TRIAL AND ERROR

Many problems which are normally solved by mathematical methods may be solved by 'trial and error' methods using an electronic calculator. The speed with which answers can be obtained to successive guesses is frequently quicker than finding the appropriate mathematical equations and solving them. Here are some examples:

(*a*) Find a number which when squared gives a four-figure answer in which the first two digits are a perfect square and also the last two digits are a perfect square. (N.B. A 'perfect square' is one which is the square of a whole number (not zero), e.g. 9 is a perfect square because it is the square of 3.)

(*b*) Find a number which equals the cube of the sum of its digits.

17 TOP FLIGHT

Civilian aircraft carry identification letters in the same way that cars have number plates. In Great Britain the aircraft are all given letters such as G–ABCD, the first letter always being G.

What is the largest number of aircraft which could be registered in Britain using this system, assuming that all the letters of the alphabet are permitted in each of the last four spaces?

18 CRICKET COMMENTARY

Bill and Ben were not the best bowlers in our village cricket team but each did a useful job of work and as we came up to the last match of the season each had got a total of ten wickets taken that season for the cost of 70 runs. There was always a bit of rivalry between them, so the situation was somewhat tense on the day of the last match, each being determined to turn in the better season's average. As the last wicket fell on the visitor's side, Bill had taken one wicket in the match for a cost of fifteen runs, whereas Ben had taken two wickets for 26 runs. Ben walked into the pavilion in jubilant mood. 'Two for twenty-six,' he said, 'That is an average of 13 runs a wicket which is better than Bill's 15 runs for the wicket he took.' But at that moment Bill walked in and announced he had won as he had got the better average for the season. This puzzled Ben who went to ask the scorer which of them had the better season's average. What did the scorer say?

19 BIG BLOW

The Treasurer of our local brass band, Bill Bluett, was counting the receipts after a concert which the band had been giving.

Collecting the silver coins together, he said 'The total amount of money here is in 50-cent, 10-cent, and 5-cent coins and is $195·50.'

Bill went on to explain that there were twice as many 10-cent coins as 5-cent coins and the number of 5-cent coins was a multiple of the number of 50-cent coins.

'But I am not going to tell you what that multiple is,' said he, 'because I want you to tell me how many of each of the coins make up the money. However, I will add that there are between thirty and forty 50-cent coins.'

Can you answer his problem?

20 TOP OF THE POPS

A magazine ran a competition in which readers had to select the ten most popular tunes of all time from a list of twenty.

How many different solutions are possible?

Suppose the magazine had asked its readers not merely to select the ten most popular tunes, but to put them in order, how many different solutions would there have been then?

21 MORE HI-FI

Bill's Hi-Fi business is thriving and he has now introduced new lines. He now stocks six different tuners, eight different record decks, five different cassette decks, seven different amplifiers, six different receivers and ten different sets of speakers.

The following combinations are needed to produce results:

> a tuner, an amplifier and a set of speakers, or
> a record deck, an amplifier and a set of speakers, or
> a cassette deck, an amplifier and a set of speakers, or
> a receiver and a set of speakers, or
> a record deck, a receiver and a set of speakers, or
> a cassette deck, a receiver and a set of speakers.

How many different combinations may a customer hear now?

22 REVERSED VIEW

Put any number into your calculator, say 1027. Now add it to the number formed by using the same figures in the reverse order; in this case it would be 7201. The answer is 8228, which reads the same forwards as it does backwards. Try other numbers.

In some cases you may have to do the process of reversing your number and adding, more than once. For example, starting with 435 you proceed:

$$435 + 534 = 969$$

But starting with 745 you proceed:

$$745 + 547 = 1292$$
$$1292 + 2921 = 4213$$
$$4213 + 3124 = 7337$$

It is impossible to prove, with a calculator, that this process always works, but it is interesting to try a large number of different starting numbers to see what happens.

Finally, see the comment in the *Hints and Answers*.

23 BRIBERY

The pirates had located the treasure on the island and overnight it was being guarded by four men but Sly Sam got past the guard and stole a number of doubloons which he proposed to smuggle onto the ship and use for his own purpose.

But on his way back he met the first guard and had to bribe him by giving him half the doubloons he carried and one doubloon more. Fortune was not on his side because he had not got much further when he met the next guard and to buy his silence he had to give him half the remaining doubloons and one doubloon more. He then met the third guard and had to bribe him by giving him half the remaining doubloons and one doubloon more. Finally the same thing happened with the fourth guard.

When Sly Sam got back to the ship he found he had only one doubloon remaining. How many doubloons did he take initially?

24 ACROSS THE CHESSBOARD

There are sixty-four squares on a chessboard, arranged in eight rows containing eight squares in each row. Suppose you put a chess-piece in the top left-hand corner and are allowed to move, one square at a time, either to the right, or down, or diagonally to the right and down, from the square on which the piece is standing. How many different routes may you use to reach the square in the diagonally opposite corner of the board, assuming, of course, that you may not move backwards?

25 ALL THE DIGITS

Your calculator keys have ten digits on them, 0, 1, 2, 3, 4, 5, 6, 7, 8, 9, and you are required to choose combinations of them together with the multiplication sign key to produce two answers to two multiplication problems. There is one condition only: you must use each digit once and once only in the two problems. For example, you may make two problems which are:

3210×4 and 5678×9

It is possible to arrange the digits in the two problems so that they both have the same answer. There are a number of different ways of doing this, so can you find the arrangements which give:

(a) the smallest possible answer
(b) the largest possible answer.

26 RECORDING SESSION

A pop group records ten numbers which are to be incorporated into a long-playing record. Assuming the A-side is always played before the B-side, in how many different ways may the order of these ten tracks be put on the record?

Next, the recording manager decides it doesn't really matter which numbers are on the B-side or in what order, but he is concerned about the choice and order of the numbers on the A-side. How many different arrangements can he make of the five numbers on the A side, assuming any of the ten may be chosen.

Finally, suppose the recording manager was not particularly fussy about the *order* of the numbers on the A-side but just wished to choose five numbers out of the ten, in how many ways could he do this?

27 SHIPWRECK

After a shipwreck, four sailors found themselves adrift on a raft together with the ship's cat. The only food they had was a box of biscuits which they agreed they would save for as long as possible. After one night at sea, however, one of the sailors awoke and decided that he would cheat by eating his share. He counted out the biscuits into four equal piles, found there was one biscuit over which he gave to the cat, ate his own share and put the other three piles back into the box.

On the second night another sailor awoke and, feeling hungry, he decided to eat his share. Since he was unaware of the previous night's happenings, he shared out the biscuits into four equal piles, found there was one biscuit over which he gave to the cat, then ate his share. He put the other three piles back into the box and did not tell his fellow sailors. On the third night the same thing happened with a third sailor; he divided the remaining biscuits into four equal piles, ate those in one pile and gave the odd biscuit over to the cat. The remaining biscuits were returned to the box.

On the fourth day, land was sighted. With shouts of joy, the men decided they would eat the biscuits, but the three guilty ones remained silent about their earlier deeds and so the biscuits were shared equally into four piles, and each man ate his share.

What is the least number of biscuits which could have been in the box initially?

28 TRAIN JOURNEY

A number of passengers board a train at its starting point. At the first stop a third of these alight and 40 new passengers get on.

At the next stop, a quarter of those on board get off and 52 passengers get on.

At the third stop a fifth of the passengers get off the train and 35 new passengers board it.

The next stop was the terminus and 163 passengers arrive there. How many passengers were on the train initially?

29 A 'POWER'FUL PROBLEM

A certain number is squared (i.e. multiplied by itself) and that result is then squared. Finally the answer is multiplied by the certain number.

The result of all this is a seven-digit number which ends in seven.

Can you find the certain number?

30 BIRD FLIGHT

Two ships are heading towards each other, one travelling at 20 knots and the other at 16 knots. When they are 63 nautical miles apart a sea-bird leaves the deck of one ship and flies towards the other ship at a constant speed of 28 knots. As soon as it reaches the second ship it turns round and flies back to the first ship, still flying at the same speed.

On reaching the first ship again, it immediately turns and flies back to the second ship and so on until the two ships pass very close to each other.

Can you find the total distance that the bird has flown?

(For the uninitiated, a knot is a speed of one nautical mile per hour.)

31 REPETITION

My mathematical girl friend, Ida Nancer, saw me using my calculator.

'You know what a prime number is, don't you?' she asked.

'Well,' said I, 'a prime is usually defined as a number which will only divide by itself and one, such as 5 and 7. But not 6, since that divides by 3 and 2 as well as 6 and one.'

'Good! Now enter any three-figure number into your calculator display,' she said.

I did and told her it was 245.

'Now,' she said, 'can you multiply that by a prime number, then multiply that answer by another prime number, and finally multiply that result by another prime number, so that you end up with an answer 245245?'

Clearly it works with any three-figure number you choose, but can you solve it?

32 GROWTH RATE

Many items, such as population, plants, money invested, etc., grow by a certain amount over each specified period, usually expressed as a percentage.

Of such items, which is increasing more rapidly, one with an increase of one per cent each week of its value at the start of the week, or one with an increase of sixty-two per cent per year?

√33 CROSS NUMBER PUZZLE

Clues across

A 13⁴

E A two-digit number, both digits even and the first twice the last

G The largest square number less than 300

H The middle two digits are the product of the outer two

J A jumbled *H* across

K A product of 17

L Consecutive digits

M A square number

O *K* across reversed

Q From 11 times *E* across take *M* across

Clues down

A The sum of the digits is the same as their product

B The last two digits is the sum of the first two

C 95×625

D Enter the numbers on the bottom row of your calculator keyboard, add the second row, then add the top row

E Eight three times

F $(2 \times 4 \times 6 \times 8) + 9^2$

N The tens digit is twice the units digit and the whole number is 301 times the number represented by the tens and units digits.

24

Clues across
- S (S across) × (Y across) = V across
- U A across divided by 13
- V See S across
- W Five to seven
- Y See S across
- Z Factorial 8 (see hint to Puzzle No. 6)

Clues down
- P The even digits
- R $71^2 - 17^2$
- S A product of 89 greater than 500
- T 29898 divided by 99
- X The number of unshaded squares in this puzzle

34 SUPERMARKET DISPLAY

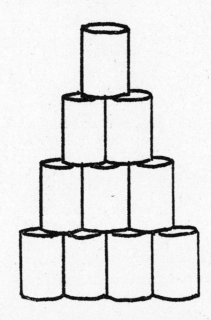

Harry Floggit has a small supermarket in our town and spends a great deal of time arranging displays of his wares in the most attractive ways.

One thing he is fond of doing is arranging piles of canned foods. Often he arranges these in a vertical triangular shape by putting a row of cans in a straight line at the bottom, then a row standing on these but with one can less than in the bottom row. The next row up contains one less can than the row it is standing on and so on until he reaches the very top row which, of course, contains only one can.

One day as I walked in he had just put a row of twelve baked bean cans on the ground. He turned to me and said, 'Do you know how many tins will be in this triangular display when I have finished it?'

Can you answer his question?

Also can you find an easy way to calculate this?

35 NEW NUMBERS

If you enter the figure 8 into your calculator and examine it closely you will find that it is made up of seven lines of light and for this digit only, all the lines are lit. For the other digits various combinations of these lines are lit to form the appropriate digit, ranging upwards from only two lines which are required to form the figure one.

By different wiring in the calculator, combinations of these lines would be possible other than those employed in forming the ten digits from 0 to 9. Of course, there would be no possibility of these other combinations being used since such 'figures' would have no meaning.

Suppose, however, that we had a change of number system from the one we use. This is not so far fetched as it may seem because it is often argued that our system of counting in tens is not the most effective. Ten will only divide exactly into halves and fifths; more common fractions than fifths are quarters, thirds and possibly eighths. But the latter create difficulties in the system of counting in tens (we call it the 'denary' system). Some people suggest that a better base for counting would be twelves, in the way in which we have twelve inches in a foot. Twelve will divide exactly by 2, 3, 4 and 6.

The difficulty with a system which counts in more than

tens is that we have insufficient digits. For example, if we were using twelve as the base of our system, we should need two extra digits, one to represent ten and one to represent eleven. Our present 10 and 11 will not do, since we need to have only one digit to represent these numbers, and so must invent new digits.

It would be similar with other number bases. Thus if we were counting in fifteens we should need five extra digits.

Now we come to the problem. How many new digits can you create which would be possible on your calculator using the basic seven lines as below?

There are restrictions, of course. The lines must be continuous or interconnected and must be the same overall height as the figure eight. This rules out combinations like two or even three horizontal lines. Also beware of slight variations of existing digits such as may be found with six or nine on different makes of calculators; some calculators show these in the form of the first two below, while others use the shape of the second two.

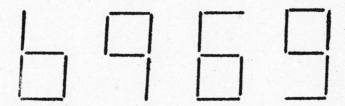

Because of these restrictions it is difficult to calculate the number of possible digits, so try drawing possible figures on paper.

36 THE LEGACY

A man arranged in his will that his four grandchildren should share the money he left when he died. To the eldest he left $123, to the second $234, to the third $345 and to the fourth $456 and directed that the remainder of his money should be divided amongst them so that the eldest received 4 parts, the next younger 3 parts, the next 2 parts and the youngest 1 part.

When the money was so divided after the man's death, it was found that the eldest received exactly $999 more than the youngest.

How much did each grandchild receive?

37 IN REVERSE

Use your calculator to square the number 201 (i.e. multiply it by itself). Make a note of your answer.

Now reverse the number 201 to obtain 102 and square this number. Compare your answer with your first answer.

Can you find some two-digit numbers which display the same property? In other words, you require a two-digit number which when squared gives a certain number and then when the two digits are reversed and the new number squared, the result has the same digits as the certain number, but in the reverse order.

38 FIRST AID

A triangular bandage is frequently used in first aid work. A piece of cotton material, initially about 40 inches square, is made into two triangular bandages by cutting across one diagonal. By folding one of these triangular pieces of material in various ways, different types of bandaging may be undertaken.

In many cases the bandage is laid flat with its longest side horizontal and the right-angle at the top. This top corner is brought over to lie on the longest side (see fig. b), then the top edge is folded over to lie on the longest side (see fig. c); this is called a broad bandage. It will be seen that the actual length of the bandage is the length of the diagonal of the original square.

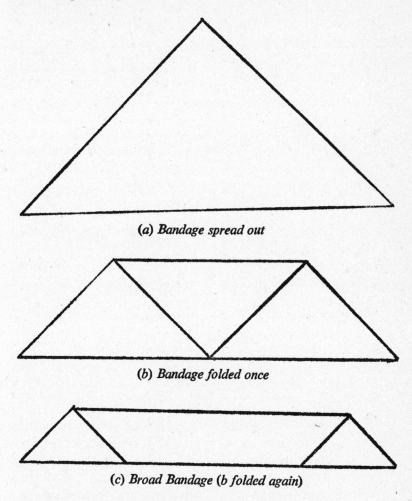

(a) Bandage spread out

(b) Bandage folded once

(c) Broad Bandage (b folded again)

Assuming that the bandage was made from a piece of material measuring 40 inches square, what is the length and width of a broad bandage?

29

39 HARRY'S SUPERMARKET

I called at Harry Floggit's supermarket again recently and found him carrying a box of cans of peaches into the store.

'Making another of your triangles, Harry?' I asked.

'I'm not certain yet,' he replied. 'With this quantity of tins I could make two triangular displays such that the bottom row of one contained exactly one more can than the bottom line of the other. On the other hand, using all these cans I could make just one triangular pile. By the way, can you tell me how many cans of peaches there are in this box?'

40 MOTORING PROBLEM

A motorist is driving up a hill on a straight road one mile long and then has to drive down the other side, which is also a straight road one mile long. If his speed up the hill is 15 m.p.h., at what speed should he drive down the other side so that his average speed for the two-mile journey shall be 30 m.p.h.?

(Warning: the answer is not 45 m.p.h.)

41 ROUNDABOUT

Two cyclists are riding around a circular track in opposite directions. The diameter of the track is 100 metres and one cyclist is travelling at a speed of 5 metres per second. The other cyclist is travelling at twice that speed.

After they pass each other, how long is it before they are at their furthest distance from each other?

How long would it take for them to be at their greatest distance from each other if they were travelling in the same direction, taking the time once more from when they were side by side?

(Take $\pi = 3 \cdot 142$ approx.)

42 METRIC MUDDLE

In the process of changing to metric measure we sometimes become confused and none more so than the apprentice at our works called Willie Gettit.

The other day he was puzzling over a piece of metal when I approached. 'Is something worrying you, Willie?' I asked.

He replied, 'I have to cut a rectangular piece of metal and I remember that its length in centimetres is five times its width in inches.'

'That's no use,' I said. 'Haven't you any other information?'

'Yes; I remember that the length of the diagonal was 2 cm longer than the length of the rectangle,' he said.

'Good,' said I, 'we can soon work out the dimensions of the rectangle now!'

What were they? (in inches)

(For information: 1 inch = $2 \cdot 54$ cm.)

43 ELECTION TIME

The candidates at an election were Ames, Benson and Collier. The votes had been cast, collected and counted and the returning officer was about to announce the result. However, he happened to be a mathematician who was fond of a puzzle and a joke. So, turning to the candidates, he said:

'The total votes cast were 9142. If we multiply the number of votes cast for Mr. Ames by the number of votes cast for Mr. Benson, the result is 859 less than the square of the number of votes cast for Mr. Collier. Finally, if we add together the votes cast for Mr. Benson and Mr. Collier, they are just 187 more than twice the number cast for Mr. Ames.'

Of course, he had to follow this by saying how many votes each candidate had received and who was elected, but can you work this out before you turn to the back of the book to find what the returning officer's final statement was?

44 PAPER SIZES

There are now standard paper sizes in use and these are denoted by reference numbers such as A3, A4, A5, etc. The point about the system is that all the sizes have their length and breadth in the same proportion and that any particular size may be cut to the next smaller size by simply cutting it into two equal parts along its length, as shown in the diagram below. Thus a sheet of A4 paper so cut would produce two sheets of A5 paper.

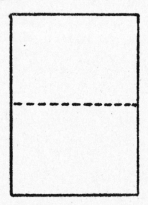

The question is in what proportion the length and width of the sheets have to be so that the system may operate?

45 EQUATORIAL JOURNEY

Two men are going to travel around the earth at the Equator. For the purpose of this problem we are going to consider the earth a perfect sphere and that the equator passes over sea or over land at sea-level. One man intends to travel around the Equator by car and boat and the other proposes to travel by aircraft flying at a constant height of 1 mile above sea-level.

Can you find how much further the man in the aircraft travels than does the man at sea-level?

Secondly, suppose we did the same problem on the moon. Again assuming it to be a perfect sphere and one of the men travelling around the Equator of the moon by lunar vehicle, while the other travels above the moon's Equator in a space vehicle maintaining a constant height of one mile above the moon's surface, how much further does the second man travel?

(Information if you require it: Earth's radius 3963 miles; moon's radius 1079 miles; $\pi = 3 \cdot 142$ approx.)

46 CAR JOURNEY

George and Mary were setting out on a journey by car and as Mary got into the car beside him, she asked George how far it was to their destination.

'I have driven there before,' said George, 'and I know the distance to the nearest tenth of a mile. I also know that if I travel at 2 more miles per hour than there are miles to travel, it will take us six minutes less than if my speed was 2 less miles per hour than there are miles to travel.'

'This is going to require pencil and paper to work out,' said Mary, producing them from her handbag.

'Quite likely,' said George, 'but your calculator will come in handy before you find the result.'

What was the distance?

47 ANTI-CALCULATOR

That famous Russian numerologist, I. Vadenuff, was talking to one of his students one day about the place of calculation in mathematical work and the way in which the electronic calculator may or may not help you in your work.

'I know it is easy to press buttons and get answers and you can do many calculations in a short space of time. However, it is not always the easiest method to use your calculator to find all the possible results to a problem; you may have to do many workings and spend much time dealing with cases which yield no result. A little mathematical working will often save much time.

'For example, suppose a positive two-digit whole number is divided by the sum of its digits, how can you find the largest and smallest possible answers without searching for all possible answers on your calculator?'

48 THE WINE MERCHANT

A wine merchant had a barrel of wine containing exactly 126 litres. From the barrel he was filling bottles of different sizes, some of them two-litre capacity, others three-litre and the remainder five-litre. He filled an exact number of bottles of each size, each bottle was completely filled and he emptied the barrel. When he had completed his task he noticed that he had filled exactly five times as many three-litre bottles as he had two-litre bottles.

Can you say how many bottles of each size he had filled?

49 AT THE NINETEENTH

The committee of our local golf club meet from time to time, of course, but apart from the committee meetings, each member comes into the club regularly but not at the same interval of time.

The Secretary comes in every day, the Captain comes in every other day, the Treasurer comes in every third day, Jones comes in every fourth day, Harrison comes in every fifth day, Walters comes in every sixth day and Jameson only comes in once a week, but always on the same day of the week.

When I was in the Clubhouse this morning I noticed that they were all there together. This set me wondering how long it would be before they were all present again, ignoring the committee meetings of course.

50 A STRANGE NUMBER

Jim was trying a mathematical trick on one of his friends at school. He had asked his friend to write down the number 12345679 (i.e. the numbers from 1 to 9 in order but missing out the 8). Jim looked at what his friend had written with a critical eye.

'Not very good writing, is it?' said he.

His friend admitted it wasn't, and when Jim asked him to choose the most badly written figure, he chose the 5.

Quickly Jim mentally multiplied this by 9 and then asked his friend to multiply the number he had written, 12345679, by 45.

A few moment's working ensued and then his friend found he was writing out the answer 555555555.

'You improved with practice,' commented Jim.

This is a well-known trick and works for any chosen number. Unfortunately we cannot transfer this to some calculator application since the answer contains nine digits which is beyond the normal eight-digit display of most calculators. You may get around it by putting a decimal point in your initial number. For example, 1234·5679 multiplied by 45 gives 55555·555, one decimal place having been lost off the end of the display.

Nevertheless, the number 12345679 is interesting in other ways. Try multiplying it by 2, then by 4, then 5, 7, 8 and 1. What do you notice about the answers in each case?

51 CHANGING SPEEDS

George, you will recall, is a rather keen motorist and a mathematician as well. I met him the other day just as he was about to set off on a journey. 'Going far?' I asked. As soon as I had said it I remembered I would not get a straight answer and sure enough George started to give me an account with which I would have to struggle to find the answer to my question.

'I intend to drive at a constant speed for the whole journey, but if I drove at 5 m.p.h. faster than that speed I would arrive six minutes earlier.'

I realized that I had to give some semblance of being able to work things out. I said, 'Ah, yes! I remember when I was at school that I was taught that speed multiplied by time equals distance, so if I multiply 5 by 6 I get 30, which is the number of miles you are travelling!'

George gave me a disgusted look. 'You have made a number of mistakes in that calculation,' he said. 'If I was going 30 miles at the second speed I would, in fact, take three minutes longer to do the journey than it would take me to do the actual journey at that speed. And now you should be able to tell me how many miles I am going to travel.'

Can you help me to retrieve my reputation, as far as George is concerned, by finding the mileage?

52 ANOTHER TRAIN JOURNEY

A train left a certain station with a number of passengers on board. At the first stop, a quarter of the passengers got off and 17 got on. At the next stop a fifth of the passengers then on the train alighted and 20 got on. When the train arrived at the next station one-eighth of those on board alighted and only one passenger boarded the train.

The next stop was the terminus and all the passengers alighted but the number of passengers getting off was not more than there were on the train when it started its journey.

How many were on the train initially?

53 A LADDER PROBLEM

A ladder, ten feet long, stands with its foot on horizontal ground and its top touching a vertical wall. In the corner between the ground and the wall is a box with a square section measuring three feet by three feet. The ladder just touches the corner of the box (see diagram).

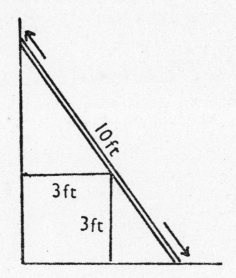

How far up the wall does the ladder reach?

54 ANOTHER LEGACY

A man left his money to be shared amongst his nephews according to a certain rule that gave most money to his favourite nephew and least money to the one he favoured least.

To Alan, the eldest, he directed that there should be given $1500, then a fifth of what remained. Next Bill would receive $1000 from what was left and a quarter of the amount remaining after deducting his $1000.

Cyril was the next nephew and he was to have $500 and then get a half of what remained after deducting his $500. Finally, Dennis was to have what was left.

When the money had been so divided, Dennis found that he had received $7007·75.

Who was the favourite nephew? And how much did he receive?

55 THE PROCESSION

Last Saturday I went to watch a procession through the centre of our town. As I took up a good position to see the passing of the procession, my friend Peter came along wearing an official-looking badge. Upon my enquiring what it was, he explained that he was one of the marshals in charge of the event. We chatted for a few moments as the procession approached, then just as the head of the procession was passing me, Peter set off to go to the end of the procession.

Some time later, he returned and said 'I have been to the end and turned round and came back immediately. The procession is $1\frac{1}{2}$ miles long and this is the half-way mark passing you now.'

It took the whole procession three-quarters of an hour to pass me and assuming that it and Peter each travelled at a constant speed, can you find Peter's speed?

Secondly, if Peter, after his brief word with me, rode on to the head of the procession at the same speed, then returned immediately to my position, can you say where the end of the procession was when he got back to my position?

56 THE INTERVIEW

That well-known firm Drabbit, Drabbit and Drabbit had advertised for someone to join their accountancy staff and Mr. Drabbit (who else?) was interviewing one of the applicants. After asking various questions which the young man answered satisfactorily, Mr. Drabbit prepared to draw the interview to a close.

'Regarding the salary,' he said to the young man, 'there are two alternatives and you would be free to choose one or the other should we offer you the appointment. The first possibility is to receive $3050 in the first year and thereafter receive an increase of $100 each year on that salary. The other alternative would be to receive $1500 in the first half-year and then an increase of $50 each half-year on that figure. Now which of these two alternatives would you choose?'

The young man thought for a moment and then gave Mr. Drabbit his answer.

Mr. Drabbit smiled and said 'You have made a good choice! And so have I, because you have got the job!'

Which alternative did the young man choose?

57 SPORTS PARADE

The two captains of the athletic teams competing in the great international competition were discussing arrangements for the grand opening parade.

'We have been told,' said the first captain, 'that each of

39

our teams has to march past the saluting base with a certain number of competitors abreast of each other in each row.'

'True,' replied the other, 'and we have to have complete rows only.'

'The only way I have been able to make the arrangement for my team is to have them marching seven abreast. Surprisingly, if I try to arrange them two abreast, three abreast, four, five or six abreast, I have always one competitor left over in each case.'

'How remarkable,' said the second captain, 'it is exactly the same with my team. There is always one person left over if they march in twos, threes, fours, fives or sixes, but the arrangement works exactly for seven in a row. And yet the number in your team is much greater than the number in my team.'

'Exactly,' replied the other captain, 'and yet the number of competitors in each team is a three-figure number.'

How many were in each team?

58 FIND THE VOLUME

I had some further trouble with Willie Gettit, the apprentice at our works, recently. Once again it was about measurement!

There were two boxes standing side by side, each one being a cube, but one being bigger than the other.

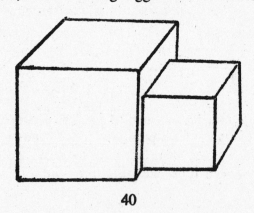

'We need to know the combined volume of these two boxes in cubic metres,' I said, handing him a tape measure and leaving him to find the answer.

Five minutes later I returned and he handed me a paper on which was written a number containing two decimal places. Being suspicious of his calculating abilities, I asked him how he had done it.

'You said you wanted the combined volume,' he said, 'so I measured along the ground from one end of one box to the far end of the other box, in metres, and that's it on the paper.'

'But,' I exclaimed, 'that is not the volume in cubic metres. It is the combined length in metres!'

So, taking the tape, I made my own measurements of each box and did the necessary calculations. When I turned round to tell him the result I suddenly noticed that my answer was the same as the one he had written on the paper. So, murmuring something about having to make an urgent phone call, I quickly left him!

Now I did notice that the length of the smaller box was exact to only one decimal place and that this box was over half the length of the larger one.

Can you say what were the dimensions of the boxes?

59 CANNY CANCELLING

When cancelling a fraction we normally divide numerator and denominator by the same number (or common factor) so that the fraction appears in a simpler form, but still has the same value. For example, $\frac{6}{8}$ is cancelled by two to obtain $\frac{3}{4}$.

However, there are certain fractions which allow a strange type of 'cancellation' by simply crossing digits out of the numerator and denominator. An example is $\frac{266}{665}$, the value of which remains the same if we simply cross out the sixes to leave the fraction $\frac{2}{5}$. (Check on your calculator that the first fraction $= 0.4$, which is also the value of the second fraction.)

Can you find some fractions with two digits in both numerator and denominator which maintain their value after similar 'cancellation'?

60 MEETING POINT

Harry and Leslie were good friends who shared a love of long distance walking. They were both surprised when they met out in the country one day and found that they were both making the same journey but in opposite directions. Harry was travelling from Newtown to Lowford and Leslie was going from Lowford to Newtown.

During their brief chat they also discovered that they had both set out on their journeys at exactly the same time but that Harry had travelled seven miles further than Leslie.

Leslie said 'Like you, I always walk at a constant speed and it will take me $17\frac{1}{2}$ hours walking to reach Newtown.'

Harry thought for a moment and then replied that it would take him $11\frac{1}{4}$ hours to get to Lowford.

Can you find how far it was from Newtown to Lowford?

61 INTERLUDE

A pair of rabbits come to live in a field, at the end of the first month they mate and a month later another pair of rabbits is born to them. From then on the original pair produce a pair of rabbits each month, while each new pair follows the same pattern of producing a pair of rabbits each month starting at the second month end. Make a note of the number of pairs of rabbits in the field each month. You will find it starts:

1, 1, 2, 3, 5 . . .

(The first four terms are the original pair and their off-spring; the fifth term includes the first pair produced by the first offspring of the original pair.)

Can you add further terms?

It reads like this:

1, 1, 2, 3, 5, 8, 13, 21 . . .

Can you now see an easy way of writing down further terms of this series? If you cannot, see the 'Hints and Answers' at the end of the book.

Now write down the first eighteen terms of this series.

Next, use your calculator to find the ratio of each number to the previous one and set your answers out in two columns as shown below:

$$\frac{1}{1} = 1 \qquad\qquad \frac{2}{1} = 2$$

$$\frac{3}{2} = 1{\cdot}5 \qquad\qquad \frac{5}{3} = 1{\cdot}6666666$$

$$\frac{8}{5} = 1{\cdot}6 \qquad\qquad \frac{13}{8} = 1{\cdot}625$$

As you proceed you will notice that the numbers in the first answer column are increasing, while those in the second answer column are decreasing. You will notice also that they are both approaching the same value. When you have done this work for 18 terms of the series, the first six figures of each of your two answers should be identical, and if you care to continue with more terms of the series and also work out more ratios, you will reach a stage when you have two equal eight-figure answers.

The series of numbers which you generated is known as the Fibonacci series. There are many interesting applications of this series, particularly in nature, e.g. the number of petals of flowers of different species; the number of spirals to be seen in the centre of the flower of a daisy.

The number to which your ratios approached is known as the Golden Ratio (or the Golden Section). It may be obtained by finding the value of $\frac{1+\sqrt{5}}{2}$. Use your calculator

43

to find $\sqrt{5}$ (see Hint to Puzzle No. 38), add 1 and divide by 2.

This ratio of $1:1\cdot618034$ . . . (the dots indicate that the number continues indefinitely) is one which was regarded as of importance in classical art and sculpture. The outside dimensions of the majority of classical paintings will be found to be in this ratio because this proportion appears to have the greatest appeal to the eye of the beholder.

The ratio is also to be found in geometry. If you draw a diagonal of a regular pentagon (i.e. five-sided figure with equal sides), the ratio of the length of the diagonal to the length of a side is the golden ratio.

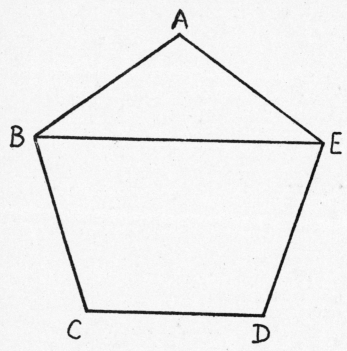

In the figure, $\dfrac{BE}{CD} = \dfrac{1+\sqrt{5}}{2}$

Turning to algebra, if we denote the golden ratio by x, it will be found that it satisfies the relation

$$x = 1 + \frac{1}{x}$$

44

To test this on your calculator, you need first to calculate

$\frac{1}{x}$ (called the reciprocal of x). Most calculators will do this

by entering the number, then pressing the division key and then the equals key (some machines require the equals key to be pressed twice). Try first with a number such as 5, to obtain an answer 0·2.

Returning to the original example, enter 1·618034, find its reciprocal and add 1. This should give your original entry, 1·618034. If there is a slight difference in the last decimal place, this is because the true value is slightly less than 1·618034.

The property you have just been demonstrating on your calculator has an interesting geometrical application. If you take a rectangle whose sides are in the ratio of the Golden Section, i.e. 1:1·618034, then cut off a square (see diagram), the remaining rectangle has its sides in the ratio of the Golden Section. You can repeat the exercise, of course, with the small rectangle which remains, continuing the process indefinitely.

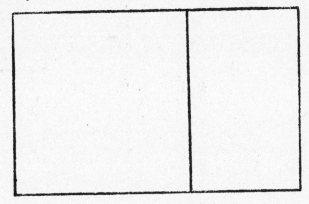

62 SALE TIME

Harry Floggitt had managed to get in a stock of certain hardware items at a reduced price and, since sales generally were a little slack, he decided to have a sale offering ten per cent off all items of hardware. Unfortunately there were certain items which he stocked on which his profit margin was so small that he didn't feel he could really allow a reduction of ten per cent in their price, so to get around this difficulty he decided to put fresh price labels on these items where the new price was ten per cent up on the previous price. Thus an article which originally had been marked 50 cents was now marked 55 cents. He then put up a notice announcing ten per cent off all marked prices.

Later in the day, however, he began to realize that things were not quite as he intended. For example, when he sold the article now marked 55 cents, he was not getting 50 cents as he hoped but only 49½ cents, since ten per cent of 55 cents is 5½ cents.

By what percentage should Harry have increased his original prices so that when he allowed ten per cent off the new price he would receive the original price?

✓ 63 CROSS NUMBER PUZZLE

Clues across

A $2^4 \times 7 \times 17 \times 59$
G 999 less than S across
I 241^3
L I could put it to one, too
M Multiples of three, but different
N A third of M across
O A multiple of 109
P Reflect on 51^2
S A quarter of a century times 290
T A multiple of 2909

Clues down

B $\frac{22}{13}$ on your calculator (and ignore the decimal point)
C A prime number
D A keyboard diagonal
E Number of seconds in a year of 365 days
F From A across subtract $2 \times 3 \times 5 \times 7$
H T across re-arranged and divisible by 10
J The last two digits are the first two, doubled and reversed
K $111 \times 23 + 103$
Q Not quite N across
R The other one (see D down)

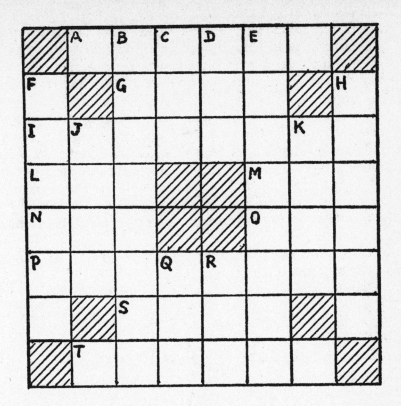

64 FULL OF BOUNCE

A rubber ball drops from a height of ten feet. After each bounce it rises to three-quarters of the height from which it fell before that bounce. What is the height of the fifth bounce? Also, on another occasion, the fourth bounce was two feet. From what height was it dropped initially?

65 KEY TO THE PROBLEM

My friend, Ida Nancer, always seems to pass by as I am using my calculator and so it was the other day.

'Ah!' she said. 'Key tapping again, I see! Perhaps you would like to do this exercise for me? You will see that there are operation keys on your calculator for add, subtract, multiply and divide and also an equals key. Then there are the number keys from 0 to 9.'

'Obviously!' said I, trying to get the better of her for a change. She gave me a withering look and went on, 'I was going to say that I want you to enter the number keys in sequence 0, 1, 2, 3, 4, 5, 6, 7, 8, 9, or in the reverse order 9, 8, 7, 6, 5, 4, 3, 2, 1, 0, and you may use any of your operation keys at any time during the process. When you finish you have to end up, in each case, with an answer of 100 in your calculator.'

Fortunately I was able to achieve the required results after a little thought and a few trials.

Can you?

66 MILK DELIVERY

Two milkmen deliver milk along our drive, working alternate weeks. There are the same number of houses on each side of the road, they are all the same size and the houses on one side face directly onto those on the other side. As usual, the odd numbers are on one side and the even numbers are on the other. There are lawns and no fences between the houses.

It appears to be necessary for the milkmen to come into our drive from one end and leave it at the other end to continue deliveries in the adjoining street. When George delivers the milk he starts at No. 1, crosses to No. 2, delivers next at No. 4, then crosses the road again to No. 3. He next delivers to No. 5, crosses the road to No. 6, then to No. 8 and so on.

However, Bill has a different method. He starts at No. 1, then goes to No. 3, No. 5 and all the odd numbers to the last house on that side, then crosses diagonally across the lawns and road to No. 2 after which he delivers to all the even numbers, finishing at the end of the row. When I mentioned this to Bill one day, he laughed and said it didn't really matter which method was used, since both routes were exactly the same length.

The distance between doorsteps is 12 metres and the distance from house to house across the road is 21·6 metres.

Can you say how many houses there are in our drive?

67 MOVING AROUND

Can you find a six-figure number, ending in 4, which is multiplied by four when the final digit is transposed to the first place?

Next, use your calculator to divide 1 by 7. The answer you obtain is only to eight figures and in the true answer the figures of the first six decimal places are repeated ad infinitum. Take these six figures, 142857, and re-arrange them with the 7 in front, but not disturbing the order, i.e. 714285. Divide this by the original number. Now try other arrangements, e.g. 857142, 428571, etc. (How many possible numbers are there?) In each case divide by the original number 142857. What do you notice?

68 FOODPACK

A manufacturer of a certain animal food packs it in bags containing 3 kilogrammes, 7 kilogrammes or 16 kilogrammes of the food. The most popular size is the 3 kg. In fact for every two bags of the 7 kg size he sells on average five 3 kg size bags. The 16 kg size is least popular, but he continues to produce it because of the demand from certain large-scale users.

He makes 190 kg of the food in one batch and packs it according to the known demand, none of the food being wasted.

Can you say how many of each size of bag he makes up from each batch of food manufactured?

69 THE BOUNCING BALL

A certain rubber ball, when dropped onto the ground rises to a height which is two-fifths of the height from which it was dropped.

If the ball is dropped initially from a height of twenty feet, how far does the ball travel before it finally comes to rest?

70 THE HANDS OF TIME

Professor I. Vadenuff drew his lecture to a close and glanced up at the clock in the lecture room.

He said 'I propose at this point to break for lunch. You will observe that the time is exactly twelve noon and the hands of the clock are such that the minute hand is exactly on top of the hour hand. We shall re-assemble here at the time when the minute hand is next on top of the hour hand. Can you say precisely in how many minutes that will be?'

71 SCHEDULED FLIGHT

An aircraft, travelling at an average speed of 550 m.p.h., leaves London for a certain destination at the same time that another aircraft starts from that destination on a journey to London, also travelling at 550 m.p.h.

Five minutes after the first aircraft leaves London, a second aircraft leaves London for the same destination, but travelling at 600 m.p.h. This aircraft overtakes the first aircraft and then 36 minutes later passes the aircraft travelling in the opposite direction, i.e. towards London.

At what time does the first aircraft touch down at its destination?

72 CHRISTMAS HAMPER

Harry Floggit was selling Christmas hampers and someone was just leaving the shop with one as I met him the other morning. After his usual greeting he said:

'I make 24 per cent profit on those hampers. But if I could have bought them for 10 per cent less than what I paid, and sold them with a profit of 28 per cent, the cost to the customer would have been 60½ cents less than what it is.'

What was the price the customer had paid for his hamper?

73 BELL RINGING

Change ringing of church bells consists of ringing the peal (i.e. the total number of bells in the belfry) in every possible order. For eight bells the full programme of changes is known as 'Bob Major' when performed in a certain order. According to the *Guinness Book of Records*, this has only once been rung ever by a team of bell ringers without using a relay team of ringers (in 1963). The total time taken was 17 hours 58 minutes.

The total number of times all the individual bells are rung is called the changes. Calculate the number of changes in ringing 'Bob Major'.

Also, using the time given above as a standard, calculate how long it takes to ring all the changes on a peal of six bells.

51

The titles applied to different types of change ringing depend in the first place on the number of bells involved in the changes. Above, the 'major' of 'Bob Major' denotes that eight bells are involved in the change ringing. The corresponding names for other numbers of bells are given below; can you say, in each case, how many changes are involved?

No. of bells	Name
4	Minimus
5	Doubles
6	Minor
7	Triples
9	Caters
10	Royal
11	Cinques
12	Maximus

74 AMATEUR DRAMATICS

I am a keen member of our local dramatic society, but I always get the jobs behind the scenes where I see neither the show nor the audience! So after the recent production, which was given on three nights, I asked our business manager whether we had had good support on each of the evenings.

'All told, 610 people saw the play,' he said. 'Fewest came on the first night, but most came on the third night and it is interesting to note that the increase in numbers of the last night over the second night was just twice the increase in numbers between the first and second nights.'

Now I know that the theatre will only accommodate an audience of 250, so I took out a pencil, paper and my calculator and set to work.

A few moments later I said 'The information you have given me gives ten possible answers to the size of audience each evening. How can I determine the correct figures?'

'I know you are interested in mathematics,' he replied, 'so I will tell you that on one of the evenings the number of people present was a perfect square, but it wasn't on the last evening.'

Can you now find the size of the audience each evening?

75 INSIDE INFORMATION

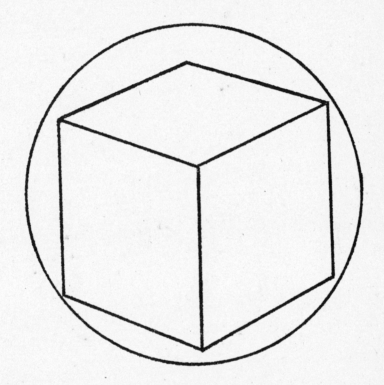

What is the length of the side of a cube which fits exactly into a sphere of diameter 10 cm. so that each of the corners of the cube touches the sphere?

If a sphere is now put inside the cube, so that this sphere touches all the sides of the cube, how much bigger is the outer sphere than the inner one?

76 THE PAINTER

The great Eastern potentate had called in a famous painter to redecorate a room of his palace. After agreeing about the scheme of decoration, the matter of payment came under discussion and the potentate asked how long the task would take. The painter said it would last twenty days.

'Very well,' said the potentate, 'now you are aware that in this country our unit of currency is the oojah, each of which is subdivided into one hundred centahs. I will offer you two methods of payment and you may choose either. In the first I will pay you one oojah on the first day, then three oojahs on the second day, five on the third day, and so on, adding a further two oojahs each day for the twenty days. Alternatively you may choose to be paid only one centah on the first day, followed by two centahs on the second day, four centahs on the third day, and so on, multiplying the amount by two for each successive day.'

Which alternative should the painter choose?

77 THE CHEMISTRY TEST

Ben Dover was not the brightest pupil in his class but was always wily enough to try to hide his results from his father, so when the latter asked Ben how he had done in the recent test in chemistry, he didn't really expect a direct answer from Ben.

'Eric and I,' said Ben, 'got the same total of marks as Alan and Derrick.'

'Tell me more!' said his father.

'Well, there are six boys in our group; Charlie and Frank in addition to those I have mentioned. Frank got twelve more marks than Charlie and Charlie got exactly half the total possible marks for the test. Also Charlie's result was 15 marks less than the average for the six of us in the group.'

'And what about the others in relation to the average?' asked his father.

Said Ben, 'Derrick had got as many marks more than Alan as Alan had got more than the average. Also, if Frank had got two marks less he would have had five-eighths of the total possible marks.'

'Very interesting,' said father, 'but one thing more. You told me about your marks added to Eric's, suppose you had added your marks to Derrick's?'

'In that case, the total would have been fourteen marks less than the total of mine with Eric's,' replied Ben.

'I see,' said his father, giving Ben a knowing look, because he had now calculated his son's marks and also his position in relation to the other five taking the test.

What were they?

78 CHANGE AROUND

In Puzzle No. 67 ('Moving Around') we dealt with a six-figure number where a figure 4 at the end was transposed to the beginning.

In this problem you are required to find a number whose first digit is 5, which when divided by 2 (and ignoring any remainder) produces an answer which has the same digits in the same order, but with the 5 transposed to the last place.

Is it possible to do a similar puzzle with any other given first digit?

79 FLIGHT PATH

An aircraft on a journey of 1800 miles encounters a strong headwind and finds that the journey takes four hours. Having reached its destination it sets out on the return journey with the wind blowing with the same speed and direction. With this tailwind the journey only takes three hours.

What would be the time for the journey if there was no wind at all?

80 A CHANGE OF HEART

A wealthy businessman decided to leave a sum of money between $200,000 and $300,000 to be shared equally between his nephews, Henry, George and Walter, and his nieces. He decided that each should receive an exact number of dollars and, being something of a mathematician, he directed that the total sum should be a perfect cube (i.e. it should be the cube of a whole number).

A year after the arrangement was made, Henry misbehaved and the man directed that he should no longer benefit from the legacy. On making the re-arrangement he was pleased to note that he did not have to alter the total amount in order that his original conditions should apply to the reduced number of beneficiaries.

Unfortunately some short time after, George also fell into disfavour and the man removed his name from the legacy, but was surprised to find that, once more, he did not need to change either the total sum or the conditions attached.

The businessman, however, was not a very hard-hearted person and within a year he was reconciled once more to both George and Henry and arranged for them to be included as beneficiaries once more. Realizing, though, that his latest action would mean that the nieces and Walter would now receive less than they had been expecting recently, he decided to increase the total amount of money available. Unfortunately he was not able to maintain what the nieces and Walter had been likely to receive before the reconciliation, but he discovered to his delight that he could arrange another figure which still conformed to all his original conditions of being a perfect cube, each person receiving an exact number of dollars and still within his two to three hundred thousand dollar limit.

He was also surprised and amused to note that if Henry and George both fell into disfavour again, they could once more be dropped from the will without any alteration in the amount or instructions about its sharing among the remaining relatives.

How many nieces were there?

81 A CONTINUED FRACTION

Can you work this out on your calculator?

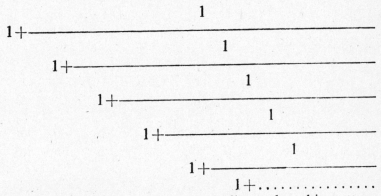

The dots at the end, of course, indicate that this pattern continues indefinitely.

When you have found the value of the fraction, can you say where you have met that value before?

82 BUSINESS INVESTMENT

Smith and Brown owned a flourishing business in which Smith had $1\frac{1}{2}$ times as much capital invested as Brown. They decided to take in a third partner and Jones was interested in the proposition.

After financial discussions it was agreed that Jones should invest $5000 in the firm and that this money should be shared between Smith and Brown in such a way that the three partners would then have equal shares in the business.

How was the $5000 shared between Smith and Brown?

83 A SECOND LADDER PROBLEM

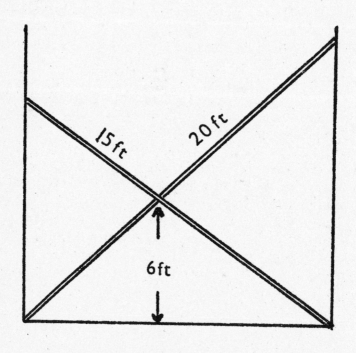

A passageway consists of a piece of horizontal ground with a vertical wall at each side. Leaning against one wall with its foot at the intersection of the opposite wall and the ground is a ladder which is 15 feet long. Almost alongside the first ladder is another ladder, 20 feet long, leaning in the opposite direction with its foot at the intersection of the ground with the other wall and its top against the opposite wall (see diagram).

It is found that the height of the level at which the two ladders cross is six feet above the ground.

Can you find the width of the passageway?

84 BIRTHDAY PARTY

Donald was having a party for his seventieth birthday and his three children and their marriage partners were present. As the meal drew to a close, Donald looked around and turning to Jean said:

'I am the oldest person present and you are the youngest and it has set me thinking about the ages of all of us. The total of the ages of the seven people around the table is 310 years, and if I exclude myself, the difference in ages of the men and women is 16 years. It is interesting to note that your age, Jean, together with Edna's age just come to the same figure as my age to-day.

'Every man is older than his wife and Tom is older than Bob by three times the number of years that Tom is older than George. Now the ages of Bob and his wife are perfect squares. Finally, my daughter is the middle child of my three children and there is exactly six years difference in age between her and my youngest son and also between her and my eldest son.'

Donald, of course, knows the relationship of the six people at the table, but given that the name of the third

59

woman at the table was Ann, can you state their relationship and their ages?

85 CANNED PYRAMIDS

It's Harry again with those confounded cans!

This time he has started arranging them in three-dimensional displays instead of arranging them in vertical triangles which are just one can deep from front to back. He now has each layer a triangular shape, starting with one can on top, then three cans in the next layer, followed by six cans in the layer below and so on.

Can you say how many cans are in the bottom layer of a pyramid of ten layers and how many cans there are altogether in a pyramid of ten layers?

Also can you see any connection between the series obtained from finding the totals in one, two, three, etc. layers and the series of triangular numbers? (See Hint to Puzzle No. 39 for triangular numbers.)

86 THREE BY THREE-PART ONE

Kathleen was busy tapping the keys on her new calculator when she looked up and said:

'If I ignore the zero and simply use the nine digits from 1 to 9, I can arrange them in groups of three so that I have a set of three numbers, each of which contains three digits. For example, one set will be 123, 456, 789. How many such sets is it possible to have, assuming that I may only use each digit once in each set?'

87 THREE BY THREE-PART TWO

Having posed her problem about the sets of three numbers each with three digits, Kathleen was not content to let the matter drop.

'Suppose,' she said, 'I add the three numbers in a set, what is the lowest possible total which I can obtain? Remember, of course, that I may only use each digit once in each set.

'I think you will find,' she went on, 'that more than one set will give you the lowest total. Can you say how many sets give that answer?

'And finally, which set of three numbers will give you the lowest answer when the three are multiplied together?'

88 A TALE OF KING ARTHUR

Now King Arthur called Merlin to him and said 'I am greatly pleased with the Knights of the Round Table. They are a fine body of men, bold, loyal and courageous; of this no-one will doubt.'

And Merlin said 'Aye, it is true! They are all worthy men!'

'But,' said King Arthur, 'there is one attribute of which I have no measure of their ability. I do not know how well they can apply their mind to a problem that will test their powers of thought.'

'You are thinking,' said Merlin, 'of the way in which they could find a solution to a problem which can be solved by thought alone.'

'That,' said the King, 'is why I seek your help. I wish you to devise a problem which I can put to the knights so that, by their answers, I shall know their wit to deal with such matters.'

'Then,' said Merlin, 'ask them this problem. Ten of the Knights of the Round Table are to dine with the King at the table which is round. Eleven chairs, exactly the same, are placed around the table and the King and his knights seat themselves. Now mark carefully the positions which they occupy at the round table, for on the next occasion the same ten knights and the King seat themselves around the table, but they must not occupy the same positions in relation to each other which they held at the first meeting. Again, a third time they are to meet together and once more must not occupy places in relation to each other which they have held on either of the previous occasions. And so it shall continue.

Now, in how many different ways could the King and his knights meet together around the table, so that on no occasion did they occupy the same positions related to each other which they had occupied before?'

'That is a good question,' said King Arthur, 'but it may well be that more than one knight shall give me the correct answer to your problem, and I am anxious to know who has the best wit among them all. Tell me one more problem which I can put to those who pass the first test, so that I may test them even further.'

'Suppose,' said Merlin, 'that a necklace of beads was to be given to Queen Guinevere on which were ten beads of different hue to represent the ten Knights of the Round Table of whom we have just spoken. And one single bead there was also, of purest white, to represent the King. Then in how many different ways could these eleven beads be arranged on the circular golden thread which formed the necklace?'

'It is good,' said King Arthur, 'The tests shall take place and we shall find the one we seek!'

Can you answer the two tests?

89 DOWN ON THE FARM

Walking down a country lane the other day, I saw my farming friend, Will Tillet, leaning on a gate and looking at some of his prize cattle.

'A fine herd you have there, Will,' I remarked.

'Aye,' he said, 'I'm very proud of 'em. Mind you, I'm proud of my goats and sheep as well. They're a fine lot!'

'You have a large farm already,' said I. 'Are you thinking of increasing your stock in the future?'

'Maybe,' said Will, 'but I would increase some more than others. Now if I added as many sheep as half the number of cattle I have, and I added as many cattle as half the number of goats I have, and I added as many goats as half the number of sheep I have, I reckon I would be right. 'Course, that would mean I had got twice as many cattle as I have now and it would mean I had got the same number of sheep as goats.'

'And how many animals would you have then?' I asked.

'I would have three hundred and ninety-six all told,' said Will.

How many of each animal has he got now?

✓ 90 DOUBLE PUZZLE

Ben Dover arrived home from school with his sister, Ann.

'It is my birthday today,' said Ben, and at school we have been doing prime numbers. Now I notice that my age today is a prime number and Ann's age is also a prime number, although there is another prime number between that representing my age and that which is hers.

'Also the next higher prime number than my age is just half my mother's age. Father's age is just twice the sum of Ann's age and mine. I am exactly the same number of years older than Ann as my father is older than my mother.

'Of course, Ann is more than five years old and if you add her age to grandpa's, the total is just two years less than the sum of my mother's and father's ages. Finally grandpa is one year older than grandma.'

Can you now solve the following Cross Number Puzzle based on the information given by Ben?

Clues across

A The square of grandma's age
E The cube of Ben's age
G The sum of all their ages.
J Twice the sum of the ages of Ben and his father
L A across minus E across
N Twelve times H down

Clues down

A The product of the ages of Ben and his mother
B Ben's age times the sum of Ann's and grandma's ages
C Grandpa's age when Ann was three
D Father's age when Ben is twenty

F Ten times the sum of the ages of the two children and their parents
H The sum of the squares of Ben's and Ann's ages
J The sum of the parents' and grandparents' ages four years ago, together with Ben's age at that time
K Ben's age in weeks
L The prime between the ages of Ben and Ann
M The sum of the ages of his parents on Ben's twenty-second birthday

91 PHONE-IN

A telephone wire a mile long is being run from the side of a building and is supported by a number of telegraph poles. It was noted that if one less pole was used, the distance between the poles would have to be increased by $2\frac{14}{15}$ yards. How many poles are being used?

92 PART BY PART

Divide the number 185058 into three parts such that half of the first part, a third of the second part and a quarter of the third part are all equal to each other.

93 THIRDS

Find a number which when diminished by a third is a third of the value of the third power of the original number.

94 THE LAST OF THE CANS

I met Harry Floggit in the street the other day. (Yes! The chap with the supermarket!)

'I've been reading your puzzle book,' he said. 'It took me some time but I managed to solve them all but for a few!'

'How many was it that you couldn't do?' I asked.

'Less than ten,' he said, grinning. 'But, you know, it set me thinking. If I had as many cans as puzzles I solved I could make either one of my triangular piles, one can deep, or I could make a pyramid of them in which every layer was a square, the sides of the squares diminishing by one can as you go up the pile!'

I thought I had heard the last of his cans, but I couldn't resist the temptation to find how many of the puzzles he had been able to do!

A NOTE ON CALCULATOR OPERATION

In some of the Hints for solution of problems details of methods of calculator operation are given. It will be found that these methods work on a majority of simple calculators. However there are variations in the methods of using certain calculators. Some have keys which perform automatically certain operations such as squaring a number or finding its reciprocal and readers with calculators having these features will ignore the methods given.

On the other hand some readers may find they need to use the processes described but do not obtain the correct result by the given method. In such cases reference should be made to the calculator instruction leaflet. Alternatively the reader may care to experiment with his calculator to find a means of obtaining the result. In such cases it is advisable to use simple numbers where the result is known (e.g. reciprocal of 4 is 0·25). Once a method has been found, the calculation of the puzzle may then be undertaken.

In the last resort, remember what the calculation really is and perform the operations in succession, e.g. the reciprocal of 4 means 1 divided by 4; the cube of 7 is 7 multiplied by 7 multiplied by 7; in each case the calculation can be entered into the calculator and the result obtained.

HINTS
AND
ANSWERS

HINTS AND ANSWERS

1 Keyboard Capers
The answer is always 1110.

The reason is that the two units figures always add up to ten, as do the hundreds figures. The same is true of the two middle digits, which are always two 5s.

2 More Keyboard Capers
The first answer is 198, the second twice this, the third three times it and the last four times 198.

The 5s may be ignored since they are subtracted. The number of hundreds in the answer comes from the difference of the two outer numbers (e.g. 6 — 4, 7 — 3, etc.) but this amount is always reduced by a similar number of units (in the case of 654 this is 200−2 = 198). The reason for the other results should now be clear.

3 Grand Total
Since the largest whole number which you can display is 99999999 and you can display every other whole number less than this, the total number must be this same figure.

Answer: 99999999. (Some readers may argue that a calculator may also give negative numbers in its display, so that we must also count all the whole numbers between −1 and −99999999. In this case, of course, the answer would be 199999998 numbers in all; we do not count 0 as a number, of course.)

4 It's an Upside Down World
No answer is needed here, but readers may care to try making up some examples for themselves involving reading the display upside down.

5 Football Fancies
There are three possible results for the first match on the list (home win, away win or draw) and for each of these

71

there are three possible results for the second match. This gives 3×3, or 9, possible ways in which the two matches may end. Proceeding in this way, the number of results of three matches is $3 \times 3 \times 3$, or using index notation, 3^3. For four matches it will be 3^4 and so on. For ten matches it is 3^{10} and for twelve is 3^{12}.

(Users with a constant multiplication facility on their calculator may use this to obtain the result quickly; usually you enter 3, press the multiplication key, then the equals key repeatedly but one time less than the power required, i.e. for power ten, press equals key nine times.)

Answers: 59049
        531441

Since there is only one correct result, your chances are 1 in 59049 and 1 in 531441 respectively.

Similar example: No. 17

6 Tom's Team
The answer is obtained from the continued multiplication $11 \times 10 \times 9 \times 8 \times 7 \times 6 \times 5 \times 4 \times 3 \times 2 \times 1$, which is easily worked out on a calculator. This continued product is called 'factorial eleven' and printed 11!

Answer: 39916800

Similar examples: No. 20, No. 73

7 Travelling Transaction
The problem may be solved by algebra, but using a calculator, start at the end. Enter 48, add 72. Divide result by 4 and add 54. Divide result by 3 and add 30. Divide result by 2.

Answer: $29

Similar examples: No. 23, No. 28

8 Re-arrangement
The answer is yes, it is always divisible by 9.

The reason is that the difference of any two of the digits is always divisible by 9. In the example given, 278165 − 187625, consider the figure ones.

In the first number this represents 100 and in the second represents 100000, the difference being 99900, which is divisible by 9. If these figures had been twos the difference would have been twice that number, if threes then three

72

times that number and so on. So for each digit, the difference gives a result divisible by 9. In some cases the result is positive, in others negative (as in the case of the ones above), but this does not alter the final result.

9   Savings Plan
The method of finding the sum may be expressed as a formula. If we have a series formed by adding the same amount to each term in order to obtain the next term, we have an *arithmetic series* and the sum of n terms is $\frac{1}{2}n(a+l)$, where a is the first term and l the last term of the series.

Answers: $136

   (*a*) $210
   (*b*) $465

Similar example: No. 34

10   Hi-Fi
For each of the eight decks, Bill can use twelve amplifiers, giving 96 possibilities. For each of these there are ten sets of speakers.

Answer: 960
Similar example: No. 21

11   Get Your Teeth Into This
Over any period, the number of teeth passing one point adjacent to any wheel is the same as those passing any point adjacent to the next wheel. We require, therefore, the smallest number of teeth into which 7, 10 and 12 divide exactly. (For those who remember their school mathematics, this is the lowest common multiple – or L.C.M. – of these numbers.)

To find this, it is necessary to express each number in its prime factors:

$$7 = 7$$
$$10 = 2 \times 5$$
$$12 = 2 \times 2 \times 3$$

The smallest number which contains all these is $2 \times 2 \times 3 \times 5 \times 7 = 420$. To find the revolutions required divide this number of teeth by 7.

Answer: 60
Similar example: No. 49

12 Lots of Squares

(i) The pattern is always the same (1, 4, 9, 6, 5, 6, 9, 4, 1) because the only figures which affect the last digits are the unit figures of the initial numbers.

(ii) Take a pair of the numbers, say, 3 and 7.

$$3^2 = 9$$
$$7^2 = (10-3)^2$$
$$= 100-60+9$$
$$= 49$$

This uses the mathematical relation $(a-b)^2 = a^2-2ab+b^2$. The result for 7^2 shows that the units digit comes from squaring 3. Similarly for the other pairs of numbers.

13 Grand Total

You have 2 parents, 2×2 or 2^2 grandparents, $2 \times 2 \times 2$ or 2^3 great-grandparents and 2^4 great-great-grandparents.

Each of them also had 2^4 great-great-grandparents.

You may care to go back into history even further and calculate the number of your predecessors at different generations. You may find that the number exceeds the world population at that time! Can you explain this?

Answers: 16, 256

Similar example: No. 17

14 Find an Easy Way

Divide 100 by three on your calculator and take the next higher whole number; multiply this by 3. Next enter 200, divide by 3, take the next lower whole number and multiply this by 3.

From this you find that the first multiple of 3 above 100 is 102 and the one immediately before 200 is 198. Their difference is 96. Of these 96 numbers between 102 and 198, a third will be multiples of three, so divide 96 by 3. This gives 32. Don't forget to add one on because you are counting *both* 102 and 198.

The second problem works in the same way.

Answers: 33
64

15 Cross Number Puzzle

Across	A 4375	E 17	F 30
	H 7569	J 1863	L 47
	M 91	O 1176 (hint; do division first)	

Down	B 31	C 7776	D 4096
	F 36	G 1145	I 5391
	K 87	N 17	

Similar examples: No. 33, No. 63

16 Trial and Error

(i) Most calculators will give squares by entering a number, say, 9, then pressing the multiplication key followed by the equals key, giving an answer in this case of 81. Returning to the present puzzle, note that a four-figure answer is required and 31^2 is less than 1000. Start by squaring 32, then 33, 34, 35, etc., until an answer meeting the required conditions is obtained.

(ii) To find the cube of a number, most calculators should be operated as for squares, but the equals key is pressed twice in succession. Start with 2^3, then find 3^3, 4^3, and so on. As each answer is obtained, mentally add up the digits. You require an answer where the sum of the digits is the same as the number you cubed, e.g. $8^3 = 512$, since $5+1+2 = 8$.

Answers: (i) $41^2 = 1681$
(ii) $17^3 = 4913$
$18^3 = 5832$

17 Top Flight

Since there are 26 letters in the alphabet, any one chosen for the first letter may be matched against 26 for the second letter, giving 26×26 combinations.

The total for the four letters is 26^4.

Answer: 456976

18 Cricket Commentary

There is a lesson in calculation here! The only way to reckon averages is to start with the basic information. Ben had 10 wickets for 70 runs initially and his latest achievement gives him 12 wickets for 96 runs overall. Your calculator will give the average as 8 runs per wicket.

Bill, however, had 11 wickets for 85 runs at the end and when this is worked out the average is 7·7272727 runs per wicket – a better average than Ben's. (N.B. that your calculator answer indicates a recurring decimal, i.e. it continues with the pattern 7272 indefinitely.)

Answer: Bill was correct and had the better average.

Similar example: No. 40

19 Big Blow

A little algebra helps in solving the problem. Suppose the number of 50-cent coins is x. Since we know that the number of 5-cent coins is a multiple of this number, we may call it kx, k and x being whole numbers, of course. Finally the number of 10-cent coins will be $2kx$.

Next we find the values of each set of coins. The value of the 50-cent coins will be $50x$ cents, of the 10-cent coins will be $20kx$ cents, and of the 5-cent coins will be $5kx$ cents.

We can now form the equation

$$50x + 20kx + 5kx = 19550$$
or
$$50x + 25kx = 19550$$

Dividing by 25

$$x(2+k) = 782$$

Since both x and $2+k$ are whole numbers, it follows that they are both factors of 782. Use your calculator to find factors and then select two which fulfil the conditions. A good hint is that the number of 50-cent coins, i.e. x, lies between 30 and 40.

Answer: 34 50-cent; 714 5-cent; 1428 10-cent.

Similar examples: No. 27, No. 52

20 Top of the Pops

Start with the second part first, where the tunes have to be arranged in order. We wish to find the total number of selections which are different.

The first tune may be any one of the 20.

The second place may be occupied by any one of the

other 19, and this applies to every one of the 20 first chosen. So there are 20×19 ways of choosing the first two places. The first three places are filled in $20 \times 19 \times 18$ ways. And so on.

So the ten tunes may be selected *in order* in $20 \times 19 \times 18 \times 17 \times 16 \times 15 \times 14 \times 13 \times 12 \times 11$ ways.

In the Hints to Puzzle No. 6, it was explained that a number such as $6 \times 5 \times 4 \times 3 \times 2 \times 1$ was called factorial six and written 6!

10! would be $10 \times 9 \times 8 \times 7 \times 6 \times 5 \times 4 \times 3 \times 2 \times 1$.

If this was multiplied onto the end of the result obtained above you will note that it would give 20!

Hence the result obtained above is $\dfrac{20!}{10!}$

We call this the permutation of 20 things taken 10 at a time and write it $_{20}P_{10}$

Now return to the first part of the question. In this case we are not concerned with the order in which the selected ten tunes appear; we are only selecting ten tunes and the same ten titles in one order is no different from those ten tunes in any other order. We call this a combination of 10 from 20 and write it $_{20}C_{10}$.

Suppose we have selected one combination of 10, then if we wished we could arrange these in order. As we saw in Puzzle No. 6, the ways of doing this is 10!

And this could be done for every combination which we had. But this would give the number of permutations of ten things taken at a time, as we have above. So it follows that:

$$_{20}C_{10} \times 10! = {}_{20}P_{10}$$

or $$_{20}C_{10} = \frac{{}_{20}P_{10}}{10!}$$

$$= \frac{20!}{10! \; 10!}$$

This last quantity is the answer to the first part of the question.

Readers may have difficulty working out some of these

quantities on their calculator because of the limitation of calculator capacity. Some manual calculation is needed to assist and careful selection can make this part easier. Thus to find $20 \times 19 \times 18 \times \ldots \times 11$, work out on your calculator $19 \times 17 \times 16 \times 14 \times 13 \times 12$. This comes to 11286912. Now 11×18 is 198 which is 2 less than 200, and 20×15 is 300.

Double your answer: 22573824
Add two 0's (i.e. multiply by 100):   2257382400
Subtract   22573824
198×11286912 = $\overline{2234808576}$

(You could use your calculator to subtract 22573824 from 57382400 and put 22 in front of the answer.)

Finally you multiply this answer by 3 and add two 0's to the end to get the required result.

Similarly to work out the answer to the first part of the puzzle, some cancellation will lighten the working.

$$\frac{20!}{10!\ 10!} = \frac{20 \times 19 \times 18 \times 17 \times 16 \times 15 \times 14 \times 13 \times 12 \times 11}{10 \times 9 \times 8 \times 7 \times 6 \times 5 \times 4 \times 3 \times 2}$$

and this cancels to $19 \times 17 \times 13 \times 4 \times 11$

Answers: 184756
        670442572800

Similar example: No. 26

21 More Hi-Fi

This puzzle works in the same way as No. 10 (Hi-Fi) except that each combination of instruments has to be worked out separately and the possible arrangements added, to obtain the grand total.

Thus the possible combinations for a tuner, amplifier and a set of speakers is $6 \times 7 \times 10 = 420$

Answer: 2170

22 Reversed View

No answer needed. Numbers which read the same forward as they do backward are called palindromic numbers.

As examples of palindromic numbers, find the values of 11^2, 11^3, 11^4, 26^2, 264^2, 307^2, 836^2.

23 Bribery

Work backwards. Enter 1 into your calculator, add 1 for the extra doubloon given to the last guard. This represents half Sly Sam's doubloons at this point, so multiply by 2. Again add 1 and multiply by 2. And so on for the four guards.

Answer: 46

Similar example: No. 28

24 Across the Chessboard

A little working on paper will show the pattern. Write down the number of ways of reaching each particular square in the position of that square.

The start of this is as follows:

```
1   1   1   1   1   1   1   1
1   3   5   7   9
1   5  13  25  41
1   7  25
1   9
1
1
1
```

The number for any square is obtained by adding together the number immediately above, the number to the left and the number diagonally to the left and above. Thus the number 41 is obtained as the sum of 25, 7 and 9.

Now continue to complete the whole square.

Answer: 48639

25 All the Digits

Answers: (a) 3485×2 and 6970×1
       (b) 915×64 and 732×80

Similar example: No. 65

26 Recording Session

For hint, see No. 20 (Hint). The answer to the first part is obtained from 10! In the second part the result is obtained from $_{10}P_5$. The last part is $_{10}C_5$.

Answers: 362880
        30240
        252

27 Shipwreck

To do the whole problem by trial and error would be tedious, so an algebraic method is the best approach. Suppose the original number of biscuits is x.

The first man removes 1 (for the cat) and eats a quarter of what is left. So he leaves $\frac{3}{4}(x-1)$.

The same is true of the second sailor except that instead of starting with x, he starts with the last result above. So he leaves $\frac{3}{4}(\frac{3}{4}(x-1)-1)$. After the third man has eaten his biscuits, there remains $\frac{3}{4}(\frac{3}{4}(\frac{3}{4}(x-1)-1)-1) = 4n$, where n is a whole number.

Simplifying

$$\frac{27}{64}x - \frac{27}{64} - \frac{9}{16} - \frac{3}{4} = 4n$$

Use your calculator to obtain these fractions as decimals and simplify to obtain the equation

$$0\cdot421875x - 1\cdot734375 = 4n$$
or
$$0\cdot421875x = 4n + 1\cdot734375$$

Now x and n are whole numbers, so if we multiply $0\cdot421875$ by whole numbers repeatedly, the answer we require must have a decimal part which reads $0\cdot734375$ (in front of the decimal, in place of the zero, there will be some whole number which is 1 more than a multiple of 4).

The method, therefore, is to put $0\cdot421875$ as constant in your calculator and multiply by 3, 4, 5, etc., successively, watching the display until you see the required decimal part appear. The method in most calculators is to enter $0\cdot421875$, press the multiplication key, then the 3 key and = key, 4 and =, 5 and =, etc.

Answer: 61

Similar example: No. 52

28 Train Journey

Work the problem backwards.

Enter 163 into your calculator and subtract 35.

This number represents four-fifths of the passengers on the train as it arrives at the third stop. To find the number, therefore, divide by 4 and multiply by 5. Alternatively, divide by $0\cdot8$, which is four-fifths.

80

Next subtract 52, divide by 3 and multiply by 4 (the reasoning is the same as before).

Deal with the first stop in the same manner.

Answer: 156

Similar example: No. 54

29 A 'Power'ful Problem

The most useful hint here is that if a number is raised to the fifth power (as it is in this problem), then the last digit of the fifth power is the same as the last digit of the original number. This reduces the number of tries you have to make with your calculator.

Answer: 17

30 Bird Flight

This is something of a catch question! There are methods of working it out which involve a great deal of calculation. However, there is a much easier approach if you think of the problem as a whole. The bird starts to fly when the ships are 63 nautical miles apart and it is flying for the total time that they take to reach the point where they pass. Since their speed of approach to each other is 36 knots, it takes them $\frac{63}{36}$ hours to cover these 63 miles. It follows that the bird is flying for the same time.

Now since the bird's speed is 28 knots, we can find the distance the bird is travelling by multiplying the time above by 28.

Answer: 49 nautical miles

Similar example: No. 55

31 Repetition

Examination of the desired result will show that the overall effect is to multiply the original number by 1001, the factors of which are the three prime numbers.

Answer: 7, 11, 13

32 Growth Rate

An increase of one per cent means that an item has increased by one-hundredth of itself in that period. Hence it increases from 1 to 1·01. Over the next period the increase is one-hundredth of this, so that at the end it becomes $1\cdot01 \times 1\cdot01$ (or $1\cdot01)^2$ of what it was originally. Over 52 weeks, therefore, the increase is $1\cdot01^{52}$.

It is easy to work this out on a calculator with a constant multiplication facility. Most calculators do this by entering a number, n say, then pressing the multiplication key, followed by the equals key. One touch of the equals key gives n^2, two touches gives n^3 and so on. So to calculate $1\cdot01^{52}$, you would enter $1\cdot01$, then press the multiplication key, followed by 51 depressions of the equals key. The increase of 62 per cent over the year would produce a final figure of $1\cdot62$, so it is necessary for the reader to find if the value of $1\cdot01^{52}$ exceeds this figure.

Answer: The weekly increase gives a faster growth.
Similar example: No. 76

33 Cross Number Puzzle

Across					
A	28561	E	84	G	289
H	3186	J	1386	K	85
L	567	M	81	O	58
Q	843	S	63	U	2197
V	2016	W	655	Y	32
Z	40320				

Down					
A	22	B	8816	C	59375
D	1368	E	888	F	465
N	18963	P	8264	R	4752
S	623	T	302	X	50

Similar example: No. 63

34 Supermarket Display

Starting at the top, the numbers in each row will be 1, 2, 3, 4, and so on to 12. Reference back to Puzzle No. 9 will show an easy way of finding the total of numbers such as these. Pairing the first and last, second and next to last, and so on, gives a total of 13 each time and there are six such pairs of numbers. Or using the formula given in the Hint to Puzzle No. 9, in this case $a = 1$, $l = 12$ and $n = 12$.

Answer: 78
Similar example: No. 39

35 New Numbers

Some of the possibilities are given below:

36 The Legacy

The lump sums add up to $1158. The number of parts in all is ten, so the share of the eldest is 0·4, that of the next is 0·3, the next 0·2 and the youngest 0·1 of the remainder. Clearly 0·4−0·1 (i.e. 0·3) is the difference in the remainder between the eldest and the youngest and the difference in the lump sum is $333.

Since $999 is the total difference, 0·3 of the remainder must be $1332. This gives the result that 0·1 is equivalent to $444 and so on for the others.

There is a trap here for the unwary, since the difference in lump sum operates in the opposite way to the difference in parts of the remainder. In other words, the youngest is $333 better off than the eldest on the lump sum alone, although the eldest has more money under the remainder arrangement. Because of this the $333 has to be added to (and not subtracted from) the $999.

Answer: Amounts received are
$1899
$1566
$1233
$ 900
Similar example: No. 54

37 In Reverse

Answers: $12 \times 12 = 144$
$21 \times 21 = 441$
$13 \times 13 = 169$
$31 \times 31 = 961$

38 First Aid

The purpose of this puzzle is to introduce the reader to two techniques which are used in later puzzles in the book.
(1) *The Theorem of Pythagoras*

In a right-angled triangle, the side which is opposite to the right-angle is called the hypotenuse and if the three sides of a right-angled triangle are measured and each of these lengths is squared, then it is found that the square of the length of the hypotenuse is equal to the sum of the squares of the lengths of the other two sides. A well-known right-angled triangle is one which has its hypotenuse of length 5 units and the other two sides of 4 and 3 units length respectively. (See diagram.) Squaring 3 gives 9 and 4 squared is 16, which when added comes to 25.

This is equal to 5 squared.

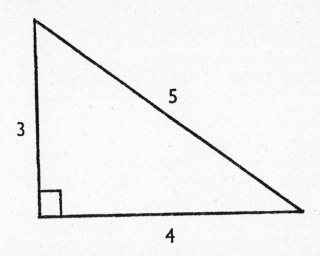

In the present example, the two shorter sides of the triangular bandage are each 40 inches. Square 40 to obtain 1600. Double this, since there are two equal sides: 3200. Now find the square root of 3200 and this will be the length of the hypotenuse, i.e. the length of the bandage when folded.

(2) *Method of finding a square root.* Some calculators have a square root key; in this case enter 3200 and press this key. In the case of other calculators it is possible to obtain square roots by successive approximation. An example will indicate the method.

Suppose we require the square root of 7. What we are looking for is a number which when multiplied by itself (i.e. squared) will give us 7. Now 2 squared is 4 and 3 squared is 9, so the number we require must lie between 2 and 3. Suppose we guess 2·5.

If this is the square root, then 2·5 times 2·5 equals 7; alternatively, if we divide 7 by 2·5 we should obtain an answer of 2·5. Doing this division on your calculator you will find that the answer is not 2·5 but 2·8. Clearly 2·5 is too small and 2·8 too large, so for a second guess we choose a number half-way between the two, i.e. 2·65, and this estimate will obviously be nearer the real value than our first one.

We now repeat the process and find that 7 divided by 2·65 is 2·6415094. Our next estimate should be between these two, and the number half-way between is 2·6457547 so we use this. Divide 7 by this and the answer is 2·6457479. You may repeat the process again, but comparing the last answer with the last estimate we see that they are identical as far as 2·6457 and this is probably as accurate as we are likely to need in the final answer. If we required greater accuracy we would continue the process, of course. Check by multiplying 2·6457 by itself to see how close to 7 is your result.

This result may be summarized into a formula, where n is the number whose square root is required, x_0 is a first approximation and x_1 is a closer approximation:

$$x_1 = \tfrac{1}{2}\left(\frac{n}{x_0} + x_0\right)$$

The order of key operation is to enter n, divide by x_0, add x_0 and finally divide by 2. The new approximation, x_1, should be written down and the whole process repeated with this new value for x_0. This is continued until the desired degree of accuracy is obtained.

Returning to our initial problem, you should now be able to find the square root of 3200 (try 56 as a first approximation).

To find the width of the bandage, remember that the height of the triangle is half the original diagonal.

Answers: 56·57 in. (to 2 decimal places)
7·07 in. (to 2 decimal places)
Similar examples: No. 42, No. 66

39 Harry's Supermarket

First refer back to Puzzle No. 34. Numbers such as these which can be represented by triangular displays are called triangular numbers. The first few triangular numbers are:

1, 3, 6, 10, 15, 21, 28, 36.

It will be seen that this series of numbers is generated by

adding firstly 2, then 3, then 4 and so on. It is possible to proceed indefinitely.

The next thing to observe is that if you take two consecutive triangular numbers and add them, you obtain a square number, thus:

$$1+3 = 4$$
$$3+6 = 9$$
$$6+10 = 16 \quad \text{and so on.}$$

In this particular example, Harry says that he can make two displays which are consecutive triangular numbers. The total number of tins in the two displays must be a square number, therefore. Next he can put all the cans together and get another triangular number. So what we are looking for is a triangular number which is also a square number. Reference to the eight triangular numbers given as an example above will reveal only one such number, 36.

You will have realized by now that a quick way to generate the triangular numbers is to put 1 into the calculator and then add successively 2, 3, 4, 5, 6, 7, etc. The result of each addition is a triangular number. See if you can find the next triangular number after 36 which is also a square.

You will find that this number (given below in the answer) is far beyond the number of cans of peaches which Harry could carry (the puzzle does say that he was 'carrying' them).

An alternative method for finding the value of any triangular number, n say, is to use the formula

$$\tfrac{1}{2}n(n+1)$$

This was explained in the Hints to Puzzle Nos. 9 and 34.

Answer: 36

> (The next number, mentioned above, is the 49th triangular number 1225)

Similar example: No. 85

40 Motoring Problem

Speeds are averages of the distance covered in a particular time and the warning given in Puzzle No. 18, applies here also. You must start from the basic information.

First calculate the time it takes to go uphill, i.e. 1 mile at 15 m.p.h. You will find that this is 4 minutes.

Next find the time needed to cover the two miles at 30 m.p.h. You will find that this is also 4 minutes.

The answer to the problem is that it cannot be done.

Similar example: No. 79

41 Roundabout

The two cyclists are at their greatest distance from each other when they are at the opposite ends of a diameter of the circle.

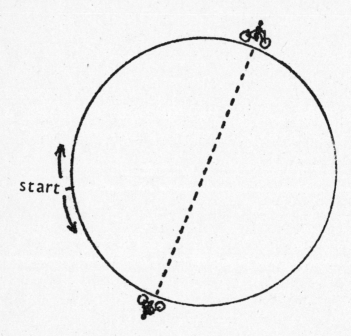

start

This means that between them they have travelled round half the circumference and since one cyclist is travelling at twice the speed of the other, one will have travelled one-sixth of the circumference and the other one-third.

Consider the cyclist who is travelling at 10 metres per second and has covered one-third of the circumference.

$$
\begin{aligned}
\text{Circumference of track} &= \pi \times \text{diameter} \\
&= 3\cdot142 \times 100 \text{ metres} \\
\text{Distance he travels} &= \tfrac{1}{3} \times 3\cdot142 \times 100 \text{ metres} \\
\text{Time to cover this distance} &= \tfrac{1}{30} \times 3\cdot142 \times 100 \text{ seconds}
\end{aligned}
$$

Part 2.

Once more it is the time taken for the cyclists to be at the opposite ends of a diameter, but this time we have to find how long it takes for half the circumference to be covered at a speed which is equivalent to the difference of their two speeds. The difference of the speeds is taken since this is the rate at which the distance between them is increasing.

Now half the circumference is $3 \cdot 142 \times 50$ metres and the difference of the two speeds is 5 metres per second.

So the time taken is $\frac{1}{5} \times 3 \cdot 142 \times 50$ seconds.

Answers: $10 \cdot 5$ seconds

$31 \cdot 4$ seconds (both answers to one decimal place)

Similar example: No. 45

42 Metric Muddle

Another problem requiring an algebraic solution.

Let the width be x inches $= 2 \cdot 54x$ cm.

Then the length is $5x$ cm and the diagonal is $5x+2$ cm.

Using Pythagoras' Theorem (see Puzzle No. 38; Hint)

$$(5x)^2+(2 \cdot 54x)^2 = (5x+2)^2$$

This simplifies to

$$25x^2+6 \cdot 4516x^2 = 25x^2+20x+4$$

or $\qquad 6 \cdot 4516x^2-20x-4 = 0$

This is a quadratic equation, since it contains the second power of the unknown (i.e. x^2) and such equations may be solved by formula (proved in all elementary algebra texts). If $ax^2+bx+c = 0$ is a quadratic equation, where a, b and c are positive or negative numbers, then the solution is given by

$$x = \frac{-b \pm \sqrt{(b^2-4ac)}}{2a}$$

(N.B. the sign $\pm$ denotes $+$ or $-$, thereby giving two answers.)

In the present problem, $a = 6 \cdot 4516$, $b = -20$ and $c = -4$.

(Notice that we have to take the negative signs into account.)

Hence $\qquad x = \dfrac{20 \pm \sqrt{(400+103 \cdot 2256)}}{12 \cdot 9032}$

89

The square root of 503·2256 may be found by successive approximation (see Hint to Puzzle No. 38) and comes to 22·433 correct to three decimal places.

So $x = \dfrac{20 \pm 22·433}{12·903}$ or $\dfrac{42·433}{12·903}$ and $\dfrac{-2·433}{12·903}$

The negative result may be ignored. Hence $x = \dfrac{42·433}{12·903}$

Answer: 3·289 inches is the width and the length is 6·474 inches (answers to three decimal places)

Similar examples: No. 43, No. 46

43 Election Time

Suppose the votes cast for Ames are denoted by a, the number cast for Benson by b and the number for Collier by c. We then have three equations:

$$a+b+c = 9142 \qquad (1)$$
$$ab+859 = c^2 \qquad (2)$$
$$2a \quad\;\; = b+c-187 \qquad (3)$$

From (1), $b+c = 9142-a$, and substitute this in (3)

$$2a = 9142-a-187$$
or $\quad 3a = 8955$
$$a = 2985$$

Equation (1) now becomes

$$b+c = 6157$$

and (2) becomes

$$2985b = c^2-859$$

which gives

$$2985\,(6157-c) = c^2-859$$

Simplifying

$$c^2+2985c-18379504 = 0$$

This is a quadratic equation as in the previous puzzle (No. 42). Because of the large numbers involved it illustrates the value of using a calculator.

$$c = \frac{-2985 \pm \sqrt{(8910225 + 73518016)}}{2}$$

$$= \frac{-2985 \pm 9079}{2}$$

= 3047 (the negative value is ignored)

Answer: Ames 2985, Benson 3110, Collier 3047

44 Paper Sizes

Suppose we denote the length of a sheet by l and its width by w, then the next smaller sheet has length w and width $\frac{1}{2}l$ (see diagram to the puzzle).

Since the proportion of length to width is always the same,

$$\frac{l}{w} \qquad \frac{w}{\frac{1}{2}l}$$

or
$$0.5l^2 = w^2$$

This gives the proportion $\frac{w}{l} = \sqrt{0.5}$

Working out the square root by successive approximation (see Puzzle No. 38; Hint) we obtain 0·7071067.

The reader will note that this is only an approximate figure since the calculation may be continued to any number of decimal places. After two decimal places, and sometimes before, the answer has no real significance in everyday measurement terms such as cutting sheets of paper.

The manuscript for this book was typed on A4 size paper and the box containing the paper gave the sheet measurement as 210×297 mm. How close is this to the desired answer?

Answer: $\dfrac{\text{width}}{\text{length}} = \dfrac{1}{0 \cdot 7071}$ (to four decimal places)

The lengths on the box (see last paragraph above) give a ratio of $1 : 0 \cdot 7070707$

45 Equatorial Journey

Call the radius of the earth R miles.

The distance travelled by the man on the surface is $2\pi R$ miles.

The radius of the circle flown by the man in the aircraft is $R+1$ miles. The distance he travels is $2\pi(R+1)$ miles.

The difference between these two distances is
$$2\pi(R+1)-2\pi R, \quad \text{or} \quad 2\pi \text{ miles.}$$

It can be seen that the distance is the same whatever the value of R. Hence the answer is the same if the problem is transferred to the moon.

Answers: 6·284 miles

6·284 miles

46   Car Journey

Suppose x miles is the total distance of the journey, then the time taken at $x+2$ m.p.h. is $\dfrac{x}{x+2}$ hours. The time taken at $x-2$ m.p.h. is $\dfrac{x}{x-2}$ hours. Since the difference of these two times is $\dfrac{1}{10}$ hour,

$$\frac{x}{x-2} - \frac{x}{x+2} = \frac{1}{10}$$

So
$$x(x+2)-x(x-2) = 0\cdot1(x-2)(x+2)$$
$$x^2+2x-x^2+2x = 0\cdot1x^2-0\cdot4$$
$$0\cdot1x^2-4x-0\cdot4 = 0$$

This is a quadratic equation and is solved by the formula method as in Hint to Puzzle No. 42.

$$x = \frac{4\pm\sqrt{(16+0\cdot4^2)}}{0\cdot2}$$

$$= \frac{4\pm4\cdot0199}{0\cdot2}$$

(4·0199 is the square root of 16·16 to 4 decimal places)

This gives two values of x, one being negative is ignored.

Answer: 40·1 miles

Similar example: No. 53

47   Anti-Calculator

A little algebra is needed here and it is important to remember that when we put two digits together, such as 44, the first 4 represents ten times the second 4, whereas if we put two letters together in algebra, such as xy, it means x multiplied by y. Let T represent the ten's digit of the number and U represent the unit's digit.

Then the number itself is $10T+U$ (if you have any doubts, test by giving values to T and U). The sum of the digits is $T+U$, so the whole number divided by the sum of its digits is

$$\frac{10T+U}{T+U}$$

which simplifies to

$$10-\frac{9U}{T+U}$$

Possible values of U range from 0 to 9 and for T range from 1 to 9 (since T cannot be zero). For a minimum value T will have to be 1 to make the last fraction as large as possible; for a maximum it will have to be 9 to make the whole fraction large.

Corresponding values of U are 9 for a minimum and 0 for a maximum.

Answer: Maximum 10; minimum $\frac{19}{10}$

Similar example: No. 59

48 The Wine Merchant

Since he filled exactly five times as many three-litre bottles as two-litre bottles, we may consider these together. If there was only one two-litre bottle filled, there would be five three-litre bottles and the total wine needed to fill them would be 17 litres.

Now the other bottles filled are five-litre ones, so the wine needed for these must be a multiple of five, which means that the number representing the litres of this wine must have a last digit of 5 or 0.

Since the last digit of the litres of wine in the barrel is 6, when we have taken all the wine for the five-litre bottles, the figure representing the wine remaining must have a last digit of 6 or 1.

This means that we are looking for a multiple of 17 which ends in 6 or 1. Put 17 in your calculator and multiply by 2, 3, etc. until you have reached 126. Make a note of any answers with a last digit of 6 or 1. The only such number is 17×3.

Answer: 3 two-litre, 15 three-litre, 15 five-litre.

Similar example: No. 68

49 At the Nineteenth

What is being sought is the smallest number into which 2, 3, 4, 5, 6, 7 will divide exactly (we do not have to worry about the Secretary who attends daily. This is obtained by finding the L.C.M. (see Hint to Puzzle No. 11) of these numbers.

2, 3, 5 and 7 are prime numbers and $4 = 2 \times 2$, and $6 = 2 \times 3$ when expressed in prime factors.

So we need a number which contains all these primes a minimum number of times.

This is $2 \times 3 \times 2 \times 5 \times 7$.

Answer: 420 days

Similar example: No. 57

50 A Strange Number

Answer: For each multiplication your display shows the digits 123456789 with one digit missing – a different one each time.

51 Changing Speeds

Suppose x miles to be the distance George is travelling. Also let his constant speed be s m.p.h.

His time would be $\dfrac{x}{s}$ hours.

If he went at a speed 5 m.p.h. faster, his time would be $\dfrac{x}{s+5}$ hours.

And the difference of these is six minutes, or one-tenth of an hour.

So his first statement gives

$$\frac{x}{s} - \frac{x}{s+5} = \frac{1}{10} \qquad (1)$$

The time to go 30 miles at the second speed is $\dfrac{30}{s+5}$ hours, and this is three minutes longer than the time for the actual journey, so

$$\frac{30}{s+5} - \frac{x}{s+5} = \frac{1}{20} \qquad (2)$$

This simplifies to $600-20x = s+5$

or $\qquad\qquad 595-20x = s$ $\qquad\qquad$ (3)

Equation (1) simplifies to

$\qquad x(s+5)-xs = 0{\cdot}1\ s(s+5)$

and using (3) to eliminate s in this equation

$\qquad 5x = 0{\cdot}1\ (595-20x)(600-20x)$

or $\qquad 5x = (595-20x)(60-2x)$

$\qquad 8x^2-479x+7140 = 0$

Solving this by formula (see Hint to Puzzle No. 42)

$$x = \frac{479\pm\sqrt{(479^2-228480)}}{16}$$

This gives $x = 28$ miles or $31{\cdot}875$ miles, but substituting the last answer in (3) shows it is unsuitable (why?).

Answer: 28 miles

Similar example: No. 60

52 Another Train Journey

This problem requires an algebraic solution. Suppose n to be the number of passengers boarding the train initially, after the first stop the number on board is $\frac{3}{4}n+17$.

At the next stop only $\frac{4}{5}$ of these remain and 20 additional passengers board the train giving

$\qquad \frac{4}{5}(\frac{3}{4}n+17)+20$

At the next station $\frac{7}{8}$ remain on board and 1 more is added, giving the number at the end as

$\qquad \frac{7}{8}[\frac{4}{5}(\frac{3}{4}n+17)+20]+1$

We are not given the final number so that we cannot obtain an equation and then solve it. All we know is that this expression is equal to a whole number; let us call this number k.

Then simplify as follows:

$\qquad \frac{7}{8}[\frac{4}{5}(\frac{3}{4}n+17)+20]+1 = k$

$\qquad \frac{84}{160}x+\frac{476}{40}+\frac{70}{4}+1 = k$

$\qquad \frac{21}{40}x+30\frac{4}{10} = k$

or $\qquad 0{\cdot}525x+30{\cdot}4 = k$

It now remains to substitute values of x which will make

95

k a whole number. There will be more than one value, and equations of this type are called diophantine equations. The use of the calculator enables different values of x to be tried rapidly until a suitable value is found.

Since k is a whole number, $0.525x$ must end in $.6$.

Enter 0.525 in your calculator and multiply in turn by 2, 3, 4, 5, etc. until your calculator shows an answer ending in $.6$.

The first value of x which gives such a result is 24, but when you try this solution in the initial problem, it does not fulfil the condition that the number of passengers at the end is not greater than it was at the beginning.

Answer: The problem as stated gives a variety of answers. The smallest value is 64. However, 104, 144, 184, etc., also work.

Similar example: No. 74

53   A Ladder Problem
This is a well-known problem which calls for special treatment in the method of its solution. Let the top of the ladder be x feet above the box and the foot of the ladder be y feet from the box, as shown in the diagram.

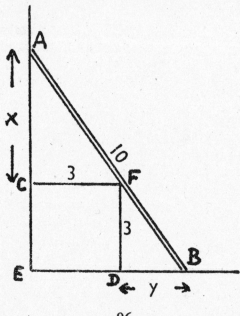

Pythagoras' Theorem has already been mentioned (see Hint to Puzzle No. 38) and we apply this to the largest triangle in the diagram to obtain:

$$(x+3)^2+(y+3)^2 = 100 \qquad (1)$$

The two similar triangles ACF and FDB have their sides in proportion, so

$$\frac{x}{3} = \frac{3}{y} \qquad (2)$$

or $\quad xy = 9$

It would seem a simple matter to substitute $\frac{9}{x}$ for y in equation (1) and so solve the problem. However if you do this you obtain

$$x^2+6x+9+\frac{81}{x^2}+\frac{54}{x}+9 = 100$$

This simplifies to

$$x^4+6x^3-82x^2+54x+81 = 0$$

This type of equation involving fourth and lower powers of x is called a bi-quadratic and such equations are difficult to solve, even with a calculator.

However, a different method of substitution produces a quadratic equation in $(x+y)$ which can first be solved (see Hint to Puzzle No. 42) and this solution may then be used to obtain a second quadratic equation which gives the final solution to the problem.

The method is as follows:

Removing brackets from equation (1)

$$x^2+6x+9+y^2+6y+9 = 100$$

From equation (2) it is clear that $2xy-18 = 0$, so if we add $2xy-18$ to the above equation, it does not alter its value, thus:

$$x^2+y^2+2xy-18+9+6x+6y+9 = 100$$
or $\quad x^2+2xy+y^2+6x+6y \qquad = 100$
i.e. $\quad (x+y)^2+6(x+y)-100 \qquad = 0$

If we put $a = x+y$, we have the quadratic equation

$$a^2+6a-100 = 0$$

Using the formula as in Hint to Puzzle No. 42,

$$a = \frac{-6 \pm \sqrt{(36+400)}}{2}$$

or $\quad a = -3 \pm \sqrt{109}$

Now the negative sign before $\sqrt{109}$ will make the whole answer negative, i.e. $x+y$ would be negative, and this is not possible in this problem. Ignoring the negative value, therefore, we have two equations

$$x+y = -3 + \sqrt{109}$$
$$xy = 9$$

Working out the value of the root gives an answer of 10·44 (to two decimal places). The second equation gives $y = \dfrac{9}{x}$ and substituting these into the first equation, we obtain

$$x + \frac{9}{x} = -3 + 10 \cdot 44$$

or $\quad x^2 - 7 \cdot 44x + 9 = 0$

Now apply the formula again to solve this quadratic equation for x. Finally add 3 feet (the side of the box) to the two values obtained to find the total distance up the wall that the ladder reaches.

Answer: 8·92 feet or 4·52 feet. (Note that whichever distance is chosen, the other length represents the distance along the ground of the foot of the ladder.)

Similar example: No. 83

54 Another Legacy

Since we know how much Dennis received and he is the last in the process of dividing the money, we can start by working from the end.

Enter into your calculator 7007·75.

Add 500. This will be Cyril's share.

Now add 7007·75 again, giving the amount after Bill had been paid. This represents three-quarters of the money before Bill had his $1000, so divide by 0·75 and add 1000.

Your answer at this stage should read 20354.

Repeating the process for Alan, divide this by 0·8 (Alan received one-fifth or 0·2) and add 1500.

This gives the total amount of the legacy as $26942·50. If

you have been noting how much each received, you can now say which nephew received most.

Answer: The favourite nephew was Cyril. He received $7507·75. Alan had $6588·50, Bill $5838·50, Dennis $7007·75

55 The Procession

The problem may be worked by algebra, but a method similar to Puzzle No. 30 yields an answer easily.

The whole procession took three-quarters of an hour to pass and Peter returned as the half-way point was passing me, so he was away for $\frac{3}{8}$ hour (or 0·375 hour). His time to reach the end of the procession must be $\frac{3}{16}$ hour (or 0·1875 hour). In that time the procession moves $\frac{3}{8}$ of a mile, so when he reaches the end, the procession has moved this distance and he has travelled $1\frac{1}{2} - \frac{3}{8} = 1\frac{1}{8}$ mile (or 1·125 mile.)

It is now possible to work out his speed since he covers 1·125 miles in 0·1875 hours. Simply divide 1·125 by 0·1875.

The second part of the question is exactly the same as the first, so the answer follows at once.

Alternatively, by calculation, to reach the front of the procession Peter has to go 0·75 miles at 4 m.p.h., since the procession's speed is 2 m.p.h. and so his overtaking speed is 4 m.p.h. Doubling this time, we find that it takes $\frac{3}{8}$ hr to get back to my position and in that time the procession moves $\frac{3}{4}$ mile.

Answers: 6 m.p.h.
He arrives back as the end of the procession is passing me.

56 The Interview

Initially there appears to be little to choose between the two alternatives, but in fact they do not give the same result.

This is demonstrated quite clearly if you list the annual incomes side by side for the two schemes, as follows:

	Scheme 1	Scheme 2	
		1st half-year	2nd half-year
1st year	3050	1500	1550
2nd year	3150	1600	1650
	and so on.		

Simple addition in this manner will show that the second scheme has the advantage over the other.

Answer: The half-yearly scheme is the better and the applicant chose this alternative.

57   Sports Parade

Start by considering the fact that the number will divide by 2, 3, 4, 5 and 6 with a remainder of 1 in each case. Forgetting the remainder for the moment, you are looking for a number which will divide exactly by these five numbers; the smallest such number is the L.C.M. (see Hint to Puzzle No. 11), which in this case is 60. This means that 60 is the smallest number into which 2, 3, 4, 5 and 6 will divide exactly, although any multiple of 60 will divide exactly by the five numbers also.

The problem is now to find a multiple of 60 which when 1 is added to it, is a multiple of 7. Use your calculator to test, e.g. enter 60, add 1, divide by 7; next enter 120, add 1, divide by 7; etc. For a solution, the answer must be a whole number. Notice the restrictions: one number is much larger than the other, but they are both three-figure numbers.

Answer: 301, 721

58   Find the Volume

Referring to the diagram in the puzzle, call the length of the side of the larger cube a and that of the smaller cube b (feet in each case).

The problem really reduces to finding suitable values for the equation

$$a+b = a^3+b^3$$

since $a+b$ is the sum of the lengths and a^3+b^3 is the sum of their volumes.

A well-known factorisation is that

$$a^3+b^3 = (a+b)(a^2-ab+b^2),$$

so in this case

$$a+b = (a+b)(a^2-ab+b^2)$$

giving

$$a^2-ab+b^2 = 1$$

100

This may be written
$$a^2-ab+\tfrac{1}{4}b^2+\tfrac{3}{4}b^2 = 1$$
or $(a-\tfrac{1}{2}b)^2 = 1-\tfrac{3}{4}b^2$ (1)

by use of the factorisation $(x-y)^2 = x^2-2xy+y^2$

Taking the square root of each side of the equation in (1)
$$a-\tfrac{1}{2}b = \pm\sqrt{(1-\tfrac{3}{4}b^2)}$$
i.e. $a = \tfrac{1}{2}b\pm\sqrt{(1-\tfrac{3}{4}b^2)}$

The negative sign will not give acceptable values.

Hence we may give values to b and use the relation
$$a = \tfrac{1}{2}b+\sqrt{(1-\tfrac{3}{4}b^2)}$$

to find corresponding values of a.

We know that b is less than a and also that it is to one decimal place only. Hence we take values of $b = 0\cdot1$, $0\cdot2$, ... $0\cdot9$ and find values of a correct to two decimal places. Thus:

$$b = 0\cdot1 \qquad a = 1\cdot05$$
$$b = 0\cdot2 \qquad a = 1\cdot08$$
$$b = 0\cdot3 \qquad a = 1\cdot12$$
and so on.

It is then necessary to check if $a+b$ is equal to a^3+b^3 when only the first two decimal places are used. For example, the first values above are unsuitable for our purpose since $a+b = 1\cdot15$ and a^3+b^3 is $1\cdot158625$, which corrected to two decimal places is $1\cdot16$.

It will be found that two values satisfy these conditions, but the statement that the size of the small box is over half that of the large box eliminates one of these.

Answer: $1\cdot12$ and $0\cdot8$ metres.

59 Canny Cancelling
The fraction will have to be of the form
$$\frac{10a+b}{10b+c} \quad (a, b \text{ and } c \text{ are digits})$$

The b in the numerator and that in the denominator would disappear under the 'cancelling', leaving $\dfrac{a}{c}$ (of course, a and c cannot be equal)

So
$$\frac{10a+b}{10b+c} = \frac{a}{c}$$

which gives $10ac+bc = 10ab+ac$

and $c = \dfrac{10ab}{9a+b}$

From this, find values of a which will give values of b and c which are whole numbers.

Answer: $\dfrac{16}{64}$ $\qquad$ $\dfrac{19}{95}$ $\qquad$ $\dfrac{26}{65}$ $\qquad$ $\dfrac{49}{98}$

60 Meeting Point

Suppose Harry and Leslie meet at a point x miles from Lowford, then it is $x+7$ miles from Newtown.

Harry's speed $= \dfrac{x}{11\frac{1}{3}}$ m.p.h.

Leslie's speed $= \dfrac{x+7}{17\frac{1}{2}}$ m.p.h.

Harry has travelled $x+7$ miles and the time he took was

$$\frac{11\frac{1}{3}(x+7)}{x} \text{ hours}$$

Leslie has travelled x miles and his time was

$$\frac{17\frac{1}{2}x}{x+7} \text{ hours}$$

Since these times are equal

$$\frac{56(x+7)}{5x} = \frac{35x}{2(x+7)}$$

Which simplifies to

$$112(x+7)^2 = 175x^2$$

and

$$9x^2-224x-784 = 0$$

This may then be solved by the formula method (see Hint to Puzzle No. 42).

Answer: 63 miles

61 Interlude

Each term of the series is the sum of the previous two terms.

62 Sale Time

When Harry adds 10 per cent onto any price, the final price is 1·10 times the original price. For example, an article costing £100 would become £100×1·10, i.e. £110. Similarly if 10 per cent is knocked off the price, the final price will be 0·90 times the original price.

In this instance we do not know how much per cent has to be added on, so suppose that the multiplying figure (corresponding to the 1·10 above) is N. Then for the article costing 50 pence Harry would mark it $50 \times N$. When he knocks off 10 per cent, he multiplies this price by 0·90 and arrives back at the original price of 50 pence, i.e.

$$50 \times N \times 0.90 = 50$$

Clearly $0.9N = 1$

Use your calculator to find $N = \dfrac{1}{0.9}$

Answer: 11·11 (to 2 decimal places)
Similar examples: No. 72

63 Cross Number Puzzle

Across					
	A	112336	G	6251	I 13997521
	L	212	M	369	N 123
	O	654	P	26011062	S 7250
	T	162904			
Down	B	16923076	C	229	D 357
	E	31536000	F	112126	H 619420
	J	3126	K	2656	Q 122
	R	159			

Similar example: No. 90

64 Full of Bounce

This example illustrates the way in which successive terms of certain series are easily generated on a calculator. Three-quarters is 0·75, so the height to which the ball rises after the first bounce will be 10×0.75.

It now drops from this height, so the height after the second bounce will be $10 \times 0.75 \times 0.75$ or 10×0.75^2.

Proceeding in the same way, the height after the fifth bounce will be 10×0.75^5.

Most calculators generate powers by entering the number, say 3, then pressing the multiplication key followed by repeated pressing of the equals key. Each touch on the equals key multiplies by 3, but remember that you had three in the calculator at the start, so the first touch of the key gives you 3^2 or 9, the second touch 3^3 or 27, and so on.

In this example, there is a slight variation. Enter 0·75 first, press the multiplication key, enter 10, then press the equals key. The first depression gives you the first power only (i.e. $10 \times 0·75$), the second depression gives you $10 \times 0·75^2$. So the fifth touch of the equals key gives you your final answer.

It is interesting to watch the answers as they appear successively because the calculator is showing you the height to which the ball rises after each bounce.

A series of numbers where each successive term is obtained by multiplying the previous term by a fixed quantity is called a geometric series and the quantity by which you are multiplying is called the common ratio for that series. In this case the common ratio is 0·75. You may find it of interest to continue the above problem to find the terms of the series beyond the fifth and notice how soon the answers become very small. After about twenty terms (i.e. nineteen bounces, since the original fall of 10 feet is the first term) it becomes almost negligible. We shall return to this matter in a later puzzle (No. 69). For the second part of this puzzle, you would enter 2 into your calculator and *divide* by 0·75 repeatedly.

Answers: 2·37 feet (to two decimal places)
6·32 feet.

Similar examples: No. 69, No. 76

65 Key to the Problem
Answer: $0+1 \div 2+3 \times 4 \times 5+6+7+8+9 = 100$
$9 \times 8+7+6+5+4+3+2+1+0 = 100$

66 Milk Delivery
Consider George first. The distance he walks *along* the street is $12(n-1)$ metres, where the number of houses on each side is represented by n (there are $n-1$ spaces between the doors).

104

However, he also walks across the road (one way or the other) n times and this distance comes to $21 \cdot 6n$ metres. His total distance is $33 \cdot 6n - 12$ metres.

Bill walks along the houses on one side (distance $12(n-1)$ metres), then across and along the houses on the other side (distance $12(n-1)$ metres). The distance across is the diagonal distance (see diagram) and this distance, by Pythagoras' Theorem is $\sqrt{\{21 \cdot 6^2 + 12^2(n-1)^2\}}$. So Bill's total distance

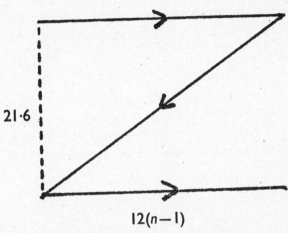

$$12(n-1)$$

is $24 (n-1) + \sqrt{(466 \cdot 56 + 144n^2 - 288n + 144)}$

Since the two distances walked by George and Bill are equal, it follows that

$$\sqrt{(466 \cdot 56 + 144n^2 - 288n + 144)} = 9 \cdot 6n + 12$$

or Squaring both sides

$$144n^2 - 288n + 610 \cdot 56 = 92 \cdot 16n^2 + 230 \cdot 4n + 144$$
$$51 \cdot 84n^2 - 518 \cdot 4n - 466 \cdot 56 = 0$$
$$n^2 - 10n + 9 = 0$$
$$(n-9)(n-1) = 0$$
$$n = 1 \text{ or } 9$$

Answer: There are 18 houses (nine on each side)

67 Moving Around

This problem is dealing with what are known as cyclic numbers. A hint to the solution of the first part of the puzzle is that the original number must start with 1, since when multiplied by 4 it gives 4. We know that the last digit is **4**, so the next to the last must be 6, since this is the result when the last figure, 4, is multiplied by 4. We now know that the number is of the form 1*XXX*64. Can you now find the other digits, here represented by *X*?

There are six possible arrangements of the number 142857 and each of them is a multiple of the original number. The number found in the first part of the puzzle only works for the movement of the one figure, 4.

Other numbers may be found displaying similar properties, but 142857 is the only number in which every possible cyclic arrangement (i.e. an arrangement preserving the order of the digits) is a multiple of the original number.

Answers: 102564

142857, 714285, 571428, 857142, 285714, 428571.

Each of these is a multiple of the first.

Similar example: No. 78

68 Foodpack

Since he sells five 3 kg bags for every two 7 kg bags, he is selling overall in units of 29 kg as far as these two sizes are concerned. Hence your answer must be based on multiples of 29. The other size is 16 kg, so it is necessary to find multiples of 29, which with multiples of 16, total 190.

One hint is that since 16 is an even number, we need look only for even multiples of 29. Work out the multiples of 16 on your calculator, taking the result from 190.

Does that answer divide by 16?

Answer: 30 three-kg bags; 12 seven-kg bags; 1 sixteen-kg bag.

69 The Bouncing Ball

Refer back to No. 64 (Full of Bounce). Once again we have a geometric series, this time with a common ratio of two-fifths or 0·4. As explained in the Hint to the earlier puzzle, constant multiplication by the common ratio soon leads to terms which are very small.

You could illustrate this in the present case by entering 20 in your calculator and multiplying repeatedly by 0·4. Since you require the total distance, you would need to add each of these results; not a very easy task unless your calculator has a memory into which each result is added as it is obtained. When the terms in your calculator become very small (or reach zero), you recall the total from the memory, double this (for the bounce up and then down) and add 20 (for the original bounce).

However, there are other ways of dealing with the summation of the series. A little mathematics is necessary.

Suppose the first term of the series is a, the common ratio is r and the sum of the series to n terms is S, then

$$S = a+ar+ar^2+ar^3+ar^4+ \ldots +ar^{n-1}$$

Multiply each side of this equation by r and write the two series below each other, like this:

$$S = a+ar+ar^2+ar^3+ \ldots +ar^{n-1}$$
$$rS = ar+ar^2+ar^3+ \ldots +ar^{n-1}+ar^n$$

Subtract the bottom row from the top:

$$S-rS = a-ar^n$$

or $\qquad S(1-r) = a(1-r^n)$

Giving

$$S = \frac{a(1-r^n)}{1-r} \tag{1}$$

Now if r is a fraction or a decimal number which is less than 1, the value of r^n becomes smaller as n becomes greater. In fact, when n is infinite, r^n becomes zero.

So for a series which goes on indefinitely (i.e. to infinity):

$$S = \frac{a}{1-r} \tag{2}$$

In theory at least, our rubber ball continues to bounce for ever, so we may find the total distance by using formula (2) above. Put $a = 20$ and $r = 0.4$. This gives 33.333333. A moment's thought will show that this is the sum of all the downward journeys. The upwards journeys will be the same except for the initial 20 feet. Hence the total distance will be $33.333333+13.333333$.

As a matter of interest, readers may care to return to formula (1) above and use it to calculate the total distance travelled in ten bounces. The difference between this and the answer for infinite bounces is only from the third decimal place, showing how quickly a geometric series of this nature can converge to a particular value.

Answer: 46·67 feet (to 2 decimal places)
Similar example: No. 76

70 The Hands of Time

The minute hand moves at a rate of 6° a minute, since it turns the full 360° in one hour. The hour hand turns at a rate of $\frac{1}{2}$° per minute.

At the end of one hour, the minute hand is back at its original point, but the hour hand is 30° further on. From the first paragraph above, it can be seen that the minute hand is now gaining on the hour hand at a rate of $5\frac{1}{2}$° per minute.

To find the time required it is necessary, therefore, to divide 30 by 5·5. Finally the 60 minutes is added to this time.

Answer: 65·45 minutes (to 2 decimal places)

71 Scheduled Flight

The speed with which the 600 m.p.h. aircraft is overtaking the first aircraft is 50 m.p.h., and when it leaves London,

the first aircraft has travelled $\frac{550}{12}$ miles. Hence the time taken

to overtake is $\frac{550}{12} \times \frac{1}{50}$ hours. This comes to 55 minutes.

The time taken for the faster aircraft to meet the one flying to London therefore, is $55 + 36 = 91$ minutes.

At that time the aircraft coming to London has travelled

$550 \times \dfrac{96}{60}$ miles. At the same time the second aircraft from

London has gone $600 \times \dfrac{91}{60}$ miles.

From these it is possible to find the total distance from London to the destination, and since the first aircraft is travelling at 550 m.p.h., division of the distance found by 550 will give the time for the journey.

Answer: 3 hr. 5 min. after take-off (to the nearest minute)

72 Christmas Hamper

The present selling price is 124 per cent of the actual cost. If Harry bought them for 10 per cent less, he would have paid 90 per cent of the actual cost for each.

His profit of 28 per cent would then make the selling price $\frac{128 \times 90}{100}$ of the actual cost.

Working this on your calculator should give 115·2.

So the difference of $60\frac{1}{2}$ cents represents $124 - 115·2$ per cent $= 8·8$ per cent.

The actual selling price is 124 per cent, so the selling price is $\frac{60·5}{8·8} \times 124$ cents.

Answer: $8.52½

73 Bell Ringing

Since the first bell to be rung may be any one of the eight, the second bell may be any one of the remaining seven and so on, the number of changes in 'Bob Major' will be 8! (See Hint to Puzzle No. 6) 8! is calculated as $8 \times 7 \times 6 \times 5 \times 4 \times 3 \times 2 \times 1$.

The answers to the other questions in the puzzle are determined in a similar fashion.

Answers: Bob Major   40320
          Time for peal of six bells is $19\frac{1}{4}$ min (approx.)

Minimus	24 changes
Doubles	120 changes
Minor	720 changes
Triples	5040 changes
Caters	362880 changes
Royal	3628800 changes
Cinques	39916800 changes
Maximus	479001600 changes

74 Amateur Dramatics

Suppose x is the number present on the first evening and d is the difference in numbers between the first and second evenings, so that the number present each evening in turn is x, $x+d$ and $x+3d$.

Adding these we know that $3x+4d = 610$.

This gives $x = \dfrac{610-4d}{3}$

$$= 203-\dfrac{4d-1}{3}$$

Use your calculator to find possible values from this, remembering that $\dfrac{4d-1}{3}$ must be a whole number. As an example, putting $d = 16$, gives $x = 203-\dfrac{64-1}{3} = 182$.

The corresponding audience figures would then be 182, 198 and 230.

List your answers and look for the figure in the first two evening attendance which is a perfect square.

Answer: 174, 196, 240

75 Inside Information
The diagonal of the cube will be the diameter of the sphere. In the diagram this is AB, which is the longest side of the triangle ABC.

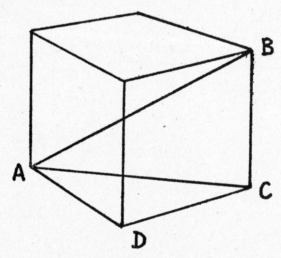

BC is one side of the cube and AC can be found from the right-angled triangle ADC.

Applying Pythagoras' Theorem (see Hint to Puzzle No. 42) to triangle ADC,

$$AC^2 = x^2 + x^2 \text{ (where } x \text{ is the length of the side}$$
$$\text{of the cube)}$$

Applying Pythagoras' Theorem to triangle ABC,

$$AB^2 = AC^2 + BC^2$$
$$= 2x^2 + x^2$$
$$= 3x^2$$

But AB is 10 cm, so $100 = 3x^2$

So $x = \dfrac{10}{\sqrt{3}}$

For the second part, the diameter of the circle clearly will be the length of the side of the cube.

Notice that 'how much bigger' is somewhat ambiguous, illustrating the care needed when dealing with solid objects. We could say either that the ratio of the diameters of the two spheres was $\dfrac{10}{5 \cdot 77} = 1 \cdot 733$, or that the ratio of their volumes was

$$\frac{10^3}{5 \cdot 77^3} = 5 \cdot 206.$$

In one case we are saying how many times longer one length is, whereas in the other case we are calculating how many times more material is contained in the sphere.

Answers: 5·77 (to 2 decimal places)
          5·206 (to 3 decimal places), but see above.

76 The Painter

Each method of payment is a series; in the case of the oojahs it is an arithmetic progression, whereas in the other case it is a geometric progression (see Puzzle No. 69). In both cases it is possible to generate successive terms on the calculator and then find the sum of them, but there are formulae to help matters.

We met a simple arithmetic progression in Puzzle No. 9 (Hint), where we were summing the series 1, 2, 3, 4, . . . 16. The sum was obtained by adding the first and last terms and multiplying by half the number of terms. The series in the present case is 1, 3, 5, 7, ... 39. The sum is $\dfrac{20}{2}(1+39)$.

Hence the amount paid is 400 oojahs.

Now refer to formula (1) in the Hint to Puzzle No. 69.

This gives the sum of the second series where a (i.e. the first term) is 1, r (the common ratio) is 2 and n is 20.

The sum is then $\dfrac{1(1-2^{20})}{1-2}$

which simplifies to $2^{20}-1$. Use your calculator to find 2^{20} (Enter 2, press $\times$ key, then press $=$ key nineteen times). Subtract 1 to obtain the answer in centahs and divide by 100 to change them to oojahs for comparison. The answer is 10485·75 oojahs.

Answer: The painter should choose the payment which starts with 1 centah per day.

77 The Chemistry Test

The problem here is sorting the information; the calculation involved is fairly simple. Suppose we denote the marks obtained by the initial letters of each individual, i.e. A is the marks Alan got, B is Ben's marks and so on. The statements may now be represented as follows:

$$
\begin{aligned}
E+B &= A+D & (1)\\
F &= C+12 & (2)\\
C &= \tfrac{1}{2}\text{ full marks} & (3)\\
C &= \text{average}-15 & (4)\\
D-A &= A-\text{average} & (5)\\
F-2 &= \tfrac{5}{8}\text{ full marks} & (6)\\
B+D &= B+E-14 & (7)
\end{aligned}
$$

From (2), (3) and (6), the difference for F gives 10 marks which is equivalent to $\tfrac{1}{8}$ of the total. Hence total possible marks is 80.

C is 40
F is 52
Average is 55

Hence the six pupils have a total of 330 marks, i.e.

$$A+B+E+D = 330-(C+F)$$

So $A+B+E+D$ $= 238$

From (1) $E+B = A+D = 119$

From (5) $D-A = A-55$

or $2A-D = 55$

$$2A-(119-A) = 55$$

$$A = 58$$

$$D = 61$$

From (7) $B+D = B+E-14$

$$D = E-14$$

$$E = 75$$

$$B = 44$$

Answer: Ben was fifth with 44 marks.

78 Change Around

The problem is easily solved by starting the division with 2 divided into 5 and putting each quotient figure into the dividend as you go along. Thus, 2 into 5 goes 2 and the dividend becomes 52; this when divided becomes 26 and the dividend is then 526.

Working on the calculator, enter 5, divide by 2, answer 2·5. Take the whole figure only and re-enter with 5 in front of it. So enter 52, divide by 2, answer 26.

Re-enter with 5 in front (i.e. 526), divide by 2, answer 263. Re-enter with 5 in front, etc. Eventually a 5 turns up in the answer and the problem is solved.

If the division is continued beyond this point, after 18 figures have been produced, the sequence starts again. A little ingenuity is needed to work out these 18 figures on a calculator, but when they have been obtained, inspection will show that the sequence contains every possible starting position, and so it is possible to perform this operation for any first digit.

The eighteen figures are 526315789473684210

Answer: 52631

79 Flight Path

The answer is *not* 3½ hours.

The reason for this is that the two journey times are

different. The problem could be worked out by algebra, but the following method on the calculator is easier.

Find the length of a four-hour journey in the 'return' direction with tailwind. (Enter 1800, divide by 3, multiply by 4.)

Now add the 1800 miles covered in the 'outward' direction with headwind. This is the total distance covered in eight hours and since there is a tailwind for four hours and a headwind for the same time, the effect of the wind cancels out.

Divide your answer by 8 and you find that the speed in still air is 525 m.p.h. Dividing this into 1800 gives the time for the journey in still air.

> Answer: 3·429 hours (to 3 dec. places) or 3 hr 26 min (to nearest minute)

80 A Change of Heart

Use your calculator to find possible perfect cubes.

The first will be 59^3 which comes to 205379. (To find a cube, enter the number, press the multiplication key and then press the equals key twice.) You will find that there are eight possible values between the limits of 200000 and 300000.

We know there are at least five beneficiaries, since it says 'nieces' and not 'neice'.

This means that from these eight numbers we are looking for one which is divisible exactly by either 3, 4 and 5, or divisible exactly by 4, 5 and 6, or by 5, 6 and 7, and so on. There is one other condition. Whichever of these triplets is chosen, the highest and lowest figures in the triple have to divide exactly into a higher number which is one of the original eight.

Investigation will show that 4, 5 and 6 divide exactly into 60^3 and that 6 and 4 divide exactly into 66^3.

> Answer: There are three nieces.

81 A Continued Fraction

Although the number is read from the top line and continues indefinitely downwards, the method of evaluation on a calculator starts at the 'bottom end' and works upwards. We may continue the working as long as we wish.

Firstly, remember that the reciprocal of a number is unity divided by that number, e.g. the reciprocal of 5 is $\frac{1}{5}$ or 0·2. To find reciprocals on a calculator, enter the number, 5 say, then press the division key followed by the equals key. If at this point your calculator shows 1 instead of 0·2, then press the equals key a second time (there is a variation with different models of calculator). Now return to the puzzle.

Imagine that the part represented by the dots has become very small; then the last fraction will be $\frac{1}{1}$ and the denominator of the previous fraction would become $\frac{1}{2}$. So start with this by putting 0·5 into your calculator. To this add 1 and find the reciprocal of the result. You now move to the next fraction by adding 1 and finding the reciprocal again. Proceed in this way (adding 1 and finding the reciprocal) ending on 'add 1' when you have arrived at a particular value which does not alter with the evaluation of further terms.

Answer: 1·618034

You have seen it before in Puzzle No. 61

82   Business Investment
Since the partners are to have equal amounts invested at the end, Smith and Brown must end up with $5000 invested each. But between them they have received $5000 from Jones and so their initial investment must have totalled $15000.

We know that Smith had $1\frac{1}{2}$ times the amount which Brown had invested, so it remains to share the $15000 in the proportion of $1\frac{1}{2}$ to 1 or 3 to 2. This means that Smith had three-fifths (or 0·6) of $15000 invested and Brown two-fifths (or 0·4).

These work out to $9000 and $6000 respectively.

Answer: Smith receives $4000 and Brown $1000.

83   A Second Ladder Problem
Like the earlier ladder problem (No. 53), this puzzle requires care in manipulation or a difficult equation is produced.

A number of similar triangles are used (see diagram):
ABC and *ECD*; *AEB* and *CEF*; *BDE* and *BCF*.

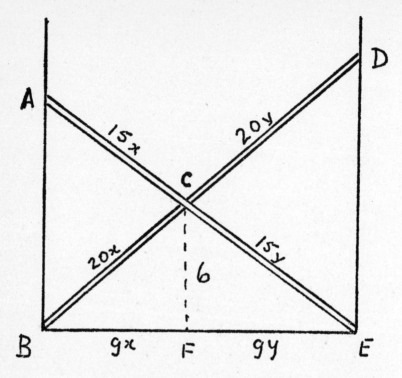

Suppose *AE* is divided at *C* in the ratio $x:y$, then $AE = 15x$, $CE = 15y$, $BC = 20x$ and $CD = 20y$. Also if the distance *BE* is called g, then $BF = gx$ and $FE = gy$.

By Pythagoras' Theorem in triangles *BCF*, *ECF*,

$$g^2x^2 = 400x^2 - 36$$
$$g^2y^2 = 225y^2 - 36$$
$$\frac{400x^2 - 36}{225y^2 - 36} = \frac{x^2}{y^2}$$

Hence $175x^2y^2 = 36(y^2 - x^2)$ (1)

By a well-known factorisation, $y^2 - x^2 = (y+x)(y-x)$ and since $x+y = 1$, it follows that $y^2 - x^2 = y - x$ and the equation (1) now becomes

$$175x^2y^2 = 36(y-x) \qquad\qquad (2)$$

Also by factorization,

$$(y-x)^2 = (y+x)^2-4xy$$
$$= 1-4xy$$

So from (2)

$$\frac{175^2x^4y^4}{36^2} = 1-4xy$$

Dividing, $23\cdot63x^4y^4+4xy = 1$

Put $a = xy$ and this equation becomes

$$23\cdot63a^4+4a = 1$$

Hence $\quad a = \dfrac{1}{23\cdot63a^3+4}$

This enables us to find a value for a by successive approximation. The method is to find a rough value of a (0·2 seems suitable) and to put this into the right-hand side of this equation and by use of the calculator find its value. The value found will be a closer approximation to the true value than 0·2 and so we repeat the process with the new value. This continues until we reach a required degree of accuracy.

Calculator operation is to enter 0·2, press multiplication key, then = key twice, multiplication key again, enter 23·63. Next press + key, then enter 4, followed by = key. (Display here should read 4·18904). Now press division key, followed by = key. Answer shown is 0·2387181, which is your closer approximation. (If your calculator is showing 1, press = key a second time; see Hint to Puzzle No. 81). Now repeat with this value in place of 0·2 and so on.

The earlier values read:

0·2
0·2387181
0·2314036
0·2329481

The reader is left to continue, the final value obtained being 0·2326833. Taking the value of a to four decimal places as 0·2327, we now have two equations:

117

$$xy = 0 \cdot 2327$$
$$x+y = 1$$

which give:

$$x^2 - x + 0 \cdot 2327 = 0$$

This may be solved by the formula method (see Hint to Puzzle No. 42) to obtain:

$$x = 0 \cdot 6315 \quad \text{or} \quad 0 \cdot 3685$$

It can be seen that one of these values is x and the other is y. Suppose y to be $0 \cdot 6315$, then in the right-angled triangle ECF,

$$CE^2 = CF^2 + FE^2$$
$$\text{So } (15 \times 0 \cdot 6315)^2 = 6^2 + (0 \cdot 6315g)^2$$
$$89 \cdot 728256 = 36 + 0 \cdot 3987922g^2$$
$$g^2 = 134 \cdot 72744$$

Answer: 11·61 feet to two decimal places.
Similar example: No. 93

84 Birthday Party

The total ages of the children and wives or husbands must be 240 and since the difference in ages of men and women is 16, the men's ages must total 128 and the women's ages total 112. Since the total ages of Edna and Jean come to 70, Ann is 42.

Turning to the men's ages, their average is between 42 and 43, and Tom is the eldest, Bob the youngest. Also Bob's age is a perfect square. Possibilities are 25 and 36, since 49 would be too old. Trying 36, multiply by 3 ($= 108$) and take this from 128.

These twenty years represent the differences in age between Bob and Tom and between Bob and George.

Tom is older than Bob by a certain number of years and George is older than Bob by two-thirds of that number of years. Therefore, we divide the 20 by 5 (to obtain 4 years), making Tom 12 years older than Bob and George 8 years older.

Now try similarly with 25 as Bob's age; the solution is not

satisfactory, you will find. Hence Bob is 36, Tom 48 and George 44.

There is six years difference between the daughter and each of her brothers, so Tom, and Bob are the sons and the daughter's age is 42. The daughter is Ann.

Since Jean is the youngest person, she must be married to Bob, because the oldest age that she can be is 35 and her age has to be a perfect square.

Hence she is 25 years old and this makes Edna 45.

Answer: Bob is married to Jean, Tom to Edna and George to Ann. Tom 48, George 44, Bob 36, Edna 45, Ann 42, Jean 25.

85 Canned Pyramids   ..

As hinted in the second part of the puzzle, the series of numbers which are obtained as totals of these cans has a connection with triangular numbers.

The triangular numbers are

1, 3, 6, 10, 15 . . .

and the totals for one, two, three, four layers as now arranged are

1, 4, 10, 20 . . .

It can be seen that just as the triangular numbers may be derived from the series of whole numbers

1, 2, 3, 4, 5 . . .

by taking the first term, then adding the first two terms, adding the first three terms and so on, of this latter series, so the new series is obtainable by adding the first one, two, three and so on, terms of the series of triangular numbers. Thus $4 = 1+3$, $10 = 1+3+6$, etc. The new series which we have obtained is known as the series of pyramid numbers. This answers the second part of the question.

The answer to the first part may be obtained by generating successive terms on your calculator until you reach the tenth term. However we found in the Hint to Puzzle No. 39 that the formula to give triangular numbers is $\frac{1}{2}n(n+1)$, so the number of cans in the tenth layer will be the tenth triangular number, which is obtained by putting $n = 10$ in this formula; the answer is 55.

Again the total number of cans may be obtained by generating the terms on your calculator and adding the results, but once more a formula gives an easier method of solution. The total number of cans in a pyramid of *n* layers is

$$\tfrac{1}{6}n(n+1)(n+2)$$

Try using the formula for the first few terms of the series to test it. Putting $n = 10$ gives the required answer.
Answer: 55
     220

86 Three by Three – Part 1

Care is needed here, since at first thought the answer would seem to be 9! The first figure of the first number in the set may be chosen in 9 ways, the next in 8, the next in 7, then the first figure of the second number may be chosen in 6 ways, and so on. But this inlcudes some duplicate arrangements since we shall have sets such as 123, 456, 789 and 456, 123, 789. These are identical since the order of the three-figure numbers in the set is unimportant.

Now the possible arrangements of the three numbers in order in the set is 3! In the above example, for instance, the first place may be occupied by either 123, 456 or 789. Similarly the second place may be occupied in two ways, making $3 \times 2 \times 1$ in all.

It is necessary, therefore, to divide the 9! by 3! This gives the answer as $9 \times 8 \times 7 \times 6 \times 5 \times 4$.
Answer: 60480

87 Three by Three – Part 2

The value of digits is greatest when they occupy the hundreds position in the number and is least when they occupy the units position. It is necessary, therefore, to put the smallest digits, 1, 2 and 3, in the hundreds position, and the 7, 8 and 9 in the units position. This leaves the tens position to be occupied by 4, 5 and 6.

One such possible arrangement is shown below and the addition is shown with the three partial sums recorded separately. Any reduction in the final total would require these partial sums to be smaller in value.

```
147
258
369
──
 24   (adding units)
150   (adding tens)
600   (adding hundreds)
──
774
──
```

Inspection of the three numbers above will show that, assuming the digits 1, 2, 3 to be fixed, 4, 5 and 6 may occupy any of these three positions, and for each of these, 7, 8 and 9 may occupy any of those three positions. In each case, therefore, there are six possible arrangements, giving 36 sets of numbers in all.

Applying similar arguments to the product problem, we are going to have once more three numbers with 1, 2 and 3 occupying the hundreds position, like this:

$$1AB \times 2CD \times 3EF$$

where the letters denote the other digits. Now A and B will be multiplied by more than 600, C and D by 300 plus, and EF by 200 plus. This and the fact that any digit occupying the position of A in the first number has ten times the value it would have if occupying the position B, should enable you to complete the numbers.

Answers: Lowest total 774
   36 sets of numbers
   Lowest product $= 147 \times 258 \times 369$
          $= 13994694$

88 A Tale of King Arthur
Since the eleven seats are exactly the same, there is no reference point until one of them is occupied.

Suppose the King enters the room first and sits in any one chair, then the chair on his right hand may be occupied by

any one of the ten knights, i.e. may be occupied in ten
different ways. Next the seat on his left hand may be occupied in any one of nine different ways, and so on. It can be
seen, therefore, that the ways of arranging the King and his
knights around the table is 10!

The necklace is a similar problem initially and once more
it would seem that the answer would be 10! However it is
possible to turn the necklace over, when we should find that
for any one of the initial arrangements, another arrangement
(the mirror image of the first) would be the same as the
initial one. As an example, in the diagram below representing
by letters a necklace with only four beads on it, the second
arrangement would be exactly the same as the first arrangement if it was turned over.

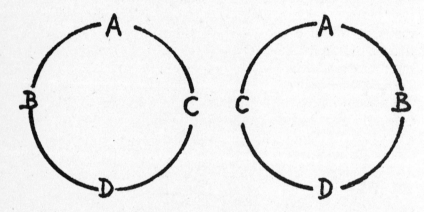

Consequently, the total possible arrangements is half of
10!

Answers: 3628800
         1814400

89 Down on the Farm

Will's first statement about adding half the number of the
various animals, coupled with his new total of 396, means
that this figure represents 1½ times his present stock. Your
calculator will show that the present stock is 396 divided by
1·5, i.e. 264.

The rest is best understood by using letters.

Suppose s, c and g represent, respectively, the number of sheep, cattle and goats which he has at present.

Then the number of goats he would have is $g+\frac{1}{2}s$ and the number of sheep he would have is $s+\frac{1}{2}c$. These would be equal, so

$$g+\frac{1}{2}s = s+\frac{1}{2}c$$

or $\qquad g = \frac{1}{2}s+\frac{1}{2}c$

This means that the goats must be one-third of the total, i.e. 88.

Finally, Will's statement that he would have twice as many cattle as he has now indicates that $c+\frac{1}{2}g = 2c$.

Hence the number of cattle is 44.

Answers: Cattle 44; goats 88; sheep 132.

90 Double Puzzle

Noting the possible school ages of Ben and his sister and the condition about prime numbers, their possible ages are 5 and 11, or 7 and 13, or 11 and 17. The first pair is discarded since Ann is over five. This gives the following possibilities from the fact that mother's age is twice the next prime and the difference between the parents' ages equals the difference between the age of Ben and Ann:

Ben	Ann	Mother	Father
13	7	34	40
17	11	38	44

The latter is ruled out by the statement that the father's age is twice the sum of the ages of Ben and Ann.

It now follows quickly from the other information that grandpa is 65 and grandma is 64.

The solution of the cross number puzzle follows.

Across

A	4096	E 2197	G 223
J	106	L 1899	N 2616

Down

A	442	B 923	C 61
D	47	F 940	H 218
J	196	K 676	L 11
M	92		

91 Phone-in

Suppose there were n poles being used, then the use of one pole less leads to an increase of $2\frac{14}{15}$ yards in each of the remaining $n-1$ spaces. So the distance between the poles initially must be $n-1$ times $2\frac{14}{15}$ yards. Since there are n of these distances

$$2\tfrac{14}{15}n(n-1) = 1760$$

This equation may be solved by algebra, but a quicker method by use of the calculator is to divide 1760 by $2\frac{14}{15}$. (The latter may be expressed as $\frac{44}{15}$, so enter 1760, divide by 44 and multiply by 15.) The result of this calculation is the product of two numbers which differ by 1, and a moment's thought should reveal the answer.

Answer: 25 poles.

92 Part by Part

Suppose half of the first part is equal to x. Then a third of the second part is also equal to x, as is a quarter of the third part.

It follows that the first part is $2x$, the second part $3x$ and the third part $4x$. Adding these we obtain $9x$. Divide the original number by 9 to find x and the parts are then obtained easily.

Answer: 41124, 61686, 82248

93 Thirds

If x is the original number, then the number diminished by x is $x-\frac{1}{3}$. A third of the value of the third power of the number is $\frac{1}{3}x^3$

Hence $\qquad\qquad \frac{1}{3}x^3 = x-\frac{1}{3}$

or $\qquad\qquad\qquad x^3 = 3x-1$

Re-arranging, $\qquad x^3-3x = -1$

$$x(x^2-3) = -1$$

$$x = \frac{1}{3-x^2}$$

This relation may be used to find successive approximations for x, as follows.

Use 1 as a first approximation and substitute in the right-hand side. (Square your approximation, take it from 3 and find the reciprocal of the result; see Hint to Puzzle No. 81) The value obtained is 0·5.

Now repeat, using 0·5 as the value of x; the answer this time is 0·3636363. Repeat, using this value as your approximation and so on.

By comparing successive answers, the required degree of accuracy is obtained. The answers for the successive aproximations are given below.

$$0·5$$
$$0·3636363$$
$$0·3487031$$
$$0·3474144$$
$$0·3473062$$
$$0·3472971$$
$$0·3472964$$
$$0·3472963$$
$$0·3472963$$

Answer: 0·3472963

94   The Last of the Cans

As we saw in earlier puzzles, the number of cans in the piles 'one can deep' is the numbers in the series of triangular numbers, 1, 3, 6, 10, etc. The number of cans in a pyramid on a square base is obtained by adding successive square numbers, thus $1+4+9+16+$ and so on.

It is necessary to compare the totals in the two cases for numbers up to 94, which is the number of this last puzzle. It will be found that two numbers are to be found which appear in both series, but one of them is eliminated by Harry's statement that the number of puzzles he could not do was less than 10.

Answer: 91

INDEX

Certain mathematical principles are involved in the solution of some of the puzzles and brief explanations of these are given at appropriate points in the text. The places where these occur are given in the index below. However, it should be pointed out that the explanations are only what is regarded as sufficient for the contents of this book and readers who require a fuller treatment should refer to appropriate elementary mathematical texts.